Encyclopedia of Environmental Issues
Policy and Activism

Encyclopedia of Environmental Issues
Policy and Activism

Editor
Craig W. Allin
Cornell College

Salem Press
A Division of EBSCO Publishing, Ipswich, Massachusetts

Cover photo:
Rainbow over Skogafoss Waterfall. (© Radius Images/Corbis)

Copyright © 2011, by Salem Press, A Division of EBSCO Publishing, Inc.
All rights in this book are reserved. No part of this work may be used or reproduced in any manner whatsoever or transmitted in any form or by any means, electronic or mechanical, including photocopy, recording, or any information storage and retrieval system, without written permission from the copyright owner except in the case of brief quotations embodied in critical articles and reviews or in the copying of images deemed to be freely licensed or in the public domain. For information address the publisher, Salem Press, at csr@salempress.com.

ISBN: 978-1-42983-669-2

Table of Contents

Contributors vii

Abbey, Edward 1
Adams, Ansel 1
Amory, Cleveland 2
Animal rights movement 3
Animal testing 5
Antienvironmentalism 7
Antinuclear movement 10
Audubon, John James 13

Back-to-the-land movement 15
Berry, Wendell 16
Bookchin, Murray 17
Borlaug, Norman 18
Brockovich, Erin 19
Brower, David 19
Brown, Lester 21
Bureau of Land Management, U.S. . . . 22
Burroughs, John 23

Ceres 24
Chipko Andolan movement 25
Cloning 26
Commoner, Barry 30
Conservation policy 31
Convention on International Trade in
    Endangered Species 36
Cousteau, Jacques 38

Darling, Jay 39

Earth First! 40
Echo Park Dam opposition 42
Ecotage 43
Ecoterrorism 44
Ehrlich, Paul R. 45
Endangered Species Act 46
Environmental law, U.S. 48
Environmentalism 52
European Green parties 55

Federal Land Policy and Management Act 56
Fish and Wildlife Act 57
Foreman, Dave 58
Fossey, Dian 59

Franklin Dam opposition 60
Friends of the Earth International 61

Gibbons, Euell 62
Gibbs, Lois 63
Gore, Al 64
Green movement and Green parties . . . 65
Greenpeace 68

Hansen, James E. 69
Hardin, Garrett 70

Inconvenient Truth, An 70
International Convention for the Regulation
    of Whaling 71
International Institute for Environment
    and Development 72
International Whaling Commission . . . 73

Land-use policy 74
League of Conservation Voters 78
Lovins, Amory 79

Maathai, Wangari 80
McToxics Campaign 81
Marine Mammal Protection Act 81
Marshall, Robert 82
Monkeywrenching 83
Muir, John 84

Nader, Ralph 86
National Audubon Society 87
Natural Resources Defense Council . . . 88
Nature writing 88

Operation Backfire 90
Osborn, Henry Fairfield, Jr. 91

People for the Ethical Treatment of
    Animals 92
Pinchot, Gifford 93
Population-control movement 94
Powell, John Wesley 97
Public opinion and the environment . . . 98

Rainforest Action Network 100

Sagebrush Rebellion 101	Watson, Paul. 114
Sale, Kirkpatrick. 103	White, Lynn Townsend, Jr. 116
SANE. 104	World Resources Institute 116
Save the Whales Campaign 105	World Trade Organization 118
Schumacher, E. F. 106	World Wilderness Congresses 119
Sea Shepherd Conservation Society 106	Worldwatch Institute 121
Silent Spring 107	
Silkwood, Karen 109	Zahniser, Howard Clinton 122
Singer, Peter. 109	
Snyder, Gary. 110	Bibliography. 124
Sun Day 111	Category Index 125
	Index. 127
Union of Concerned Scientists. 112	
U.S. Climate Action Partnership 114	

Contributors

Craig W. Allin
Cornell College

Anita Baker-Blocker
Ann Arbor, Michigan

Ruth Bamberger
Drury College

David Landis Barnhill
Guilford College

Alvin K. Benson
Utah Valley University

Massimo D. Bezoari
Huntingdon College

Cynthia A. Bily
Macomb Community College

Lakhdar Boukerrou
Florida Atlantic University

Howard Bromberg
University of Michigan

Kenneth H. Brown
Northwestern Oklahoma State University

Aubyn C. Burnside
Hickory, North Carolina

Dale F. Burnside
Lenoir-Rhyne College

Nader N. Chokr
Oslo, Norway

Thomas Clarkin
University of Texas at San Antonio

Kathryn A. Cochran
Longview Community College

Jonelle DePetro
Eastern Illinois University

Joseph Dewey
University of Pittsburgh

Gary E. Dolph
Indiana University Kokomo

Colleen M. Driscoll
Villanova University

Thomas R. Feller
Nashville, Tennessee

Soraya Ghayourmanesh
Bayside, New York

Jerry E. Green
Miami University

Wendy Halpin Hallows
Chestnut Hill College

Wendy C. Hamblet
North Carolina A&T State University

Clayton D. Harris
Middle Tennessee State University

Howard V. Hendrix
California State University, Fresno

Jane F. Hill
Bethesda, Maryland

John R. Holmes
Franciscan University of Steubenville

Louise D. Hose
Westminster College

Ronald K. Huch
University of Papua New Guinea

Raymond Pierre Hylton
Virginia Union University

Allan Jenkins
University of Nebraska at Kearney

Albert C. Jensen
Central Florida Community College

Bruce E. Johansen
University of Nebraska at Omaha

Karen N. Kähler
Pasadena, California

Karen E. Kalumuck
The Exploratorium

Robert W. Kingsolver
Kentucky Wesleyan College

Timothy Lane
Louisville, Kentucky

Eugene Larson
Los Angeles Pierce College

Thomas T. Lewis
Mount Senario College

Michael Mooradian Lupro
North Carolina A&T State University

Larry S. Luton
Eastern Washington University

Joel P. MacClellan
University of Tennessee

Robert McClenaghan
Pasadena, California

Nancy Farm Männikkö
Centers for Disease Control and Prevention

Chogollah Maroufi
California State University, Los Angeles

Charles Mortensen
Ball State University

Peter Neushul
California Institute of Technology

David L. O'Hara
Augustana College

Gordon A. Parker
University of Michigan-Dearborn

Aaron S. Pollak
Omaha, Nebraska

Oliver B. Pollak
University of Nebraska at Omaha

Charles W. Rogers
Southwestern Oklahoma State University

Robert M. Sanford
University of Southern Maine

Elizabeth D. Schafer
Loachapoka, Alabama

Alexander Scott
Pasadena, California

Amy Sisson
Houston Community College

Adam B. Smith
University of California, Berkeley

Roger Smith
Portland, Oregon

Anne Statham
University of Wisconsin-Parkside

Joan C. Stevenson
Western Washington University

Theresa L. Stowell
Adrian College

Hubert B. Stroud
Arkansas State University

Charles L. Vigue
University of New Haven

C. J. Walsh
Mote Marine Laboratory

Shawncey Webb
Taylor University

Thomas A. Wikle
Oklahoma State University

Marcie L. Wingfield
Huntingdon College

Abbey, Edward

CATEGORIES: Activism and advocacy; preservation and wilderness issues
IDENTIFICATION: American environmental activist and author
BORN: January 29, 1927; Indiana, Pennsylvania
DIED: March 14, 1989; Tucson, Arizona
SIGNIFICANCE: The originality of Abbey's ideas regarding the preservation of nature, expressed with great eloquence in his writings, helped to increase awareness of environmental issues and inspired a radical environmental movement.

Over the course of his lifetime, Edward Abbey produced twenty-one volumes of fiction, essays, speeches, and letters expressing his love for the earth, his hatred of modern technological society, and his fervent belief that development was destroying the American West. When he was twenty-one years old, having spent some time in the military and in college, he left his home in Pennsylvania to see the American West. He hitchhiked, rode trains, and walked over the mountains and through the desert. He claimed the desert as his spiritual home and lived in or near it for most of the rest of his life. He completed a master's degree at the University of New Mexico and wrote his Ph.D. thesis on anarchism and the morality of violence. During his ten-year college career, which included two years as a Fulbright Fellow at the University of Edinburgh in Scotland, he began a number of writing projects and published his first novel, *Jonathan Troy* (1954).

For fifteen years, during his thirties and forties, Abbey worked as a part-time ranger at various national parks in the American Southwest. The two years in the late 1950's that he spent at Arches National Monument (now a national park) in Utah led to his first important book, *Desert Solitaire: A Season in the Wilderness* (1968). This book combines beautiful descriptive passages, an unflinching look at the violence in nature, and a strong call for the preservation of desert habitats. Reminiscent of Henry David Thoreau's *Walden* (1854) in its ideas and its use of the natural year for its structure, *Desert Solitaire* brought Abbey national attention as an environmental writer.

In 1975 Abbey published *The Monkey Wrench Gang*, a novel about four rebels who set out to destroy the roads, bridges, and power lines that they believe are defacing the southwestern desert. This work is loosely based on the exploits of a friend of Abbey who had committed some of the acts depicted in the novel. Despite the fact that Abbey consistently maintained that he intended the book primarily as humor, it helped inspire the radical environmental group Earth First!, a group that Abbey did come to support, praising its operations although never actually joining it. In fact, Abbey never joined any political or environmental organizations, although he participated in political actions, especially those that expressed disapproval of the military or land development.

Though most of his books are set in the wilderness of the Southwest and express his deep love for such spaces, Abbey disliked being called a "nature writer." In fact, students of his work have had difficulty attaching any label to Abbey and making it stick, so idiosyncratic are his ideas and connections. When Abbey died in 1989, he left instructions that he should be buried in the desert, unembalmed, in his sleeping bag. Although this kind of burial was illegal, his friends followed his wishes.

Cynthia A. Bily

FURTHER READING

Bishop, James, Jr. *Epitaph for a Desert Anarchist: The Life and Legacy of Edward Abbey.* New York: Atheneum, 1994.
Cahalan, James M. *Edward Abbey: A Life.* Tucson: University of Arizona Press, 2001.
Pozza, David M. *Bedrock and Paradox: The Literary Landscape of Edward Abbey.* New York: Peter Lang, 2006.

Adams, Ansel

CATEGORIES: Activism and advocacy; preservation and wilderness issues
IDENTIFICATION: American photographer and environmental activist
BORN: February 20, 1902; San Francisco, California
DIED: April 22, 1984; Carmel, California
SIGNIFICANCE: Through his spectacular photographs and his advocacy, Adams helped to increase Americans' awareness of the beauty of the nation's wilderness areas and the importance of preserving that beauty.

Ansel Adams was the only son of Charles and Olive Bray Adams. Although a gifted child, he detested public schools and was educated primarily by tutors, graduating from a private school in 1917. His early interests centered on music and, after he acquired his first camera, photography. A visit to Yosemite National Park in California in 1916 sparked his interest in nature photography, and he made frequent return visits to Yosemite throughout his lifetime.

Adams was able to support himself with music and photography during his early working years. As his interest in photography increased, however, it became the dominant factor in his life. He traveled extensively throughout the American West, photographing landscapes. His first published photograph appeared in the *Sierra Club Bulletin* in 1927, and the first of his numerous photographic collections was published in the same year. In 1928 he married Virginia Rose Best in Yosemite and held his first one-person exhibit in San Francisco. The Adamses had two children: Michael, born in 1933, and Anne, born in 1935.

In 1934 Adams was elected to the board of directors of the Sierra Club, a position he held until 1971. He was a cofounder of Group f/64, an organization dedicated to the use of photography to emphasize and preserve the natural beauty of the American West.

Adams's photographs, almost exclusively in black and white, are renowned for their sharp contrast, detail, use of light and shadow, and ability to capture the beauty of their natural settings. During his lifetime, Adams often exhibited his work in museums and at universities that recognized his artistic talent. In addition to this artistry, he developed several innovative techniques for developing photographs to enhance their contrast and embellish their appearance. He published several how-to books on photography and taught workshops on how to utilize his techniques. In addition, he established his own studio to exhibit and sell his works.

Largely self-taught, Adams is recognized as one of the leaders in nature photography. He not only photographed nature but also realized the need for environmental activism and lobbied extensively for conservation measures. As early as 1936 he approached the U.S. Congress to promote the establishment of additional national parks in the western states. He was a member of President Lyndon B. Johnson's environmental task force and later met with both President Gerald Ford and President Ronald Reagan to discuss environmental issues.

Adams's work extended beyond nature photography to include architectural studies, portraits, and commercial photography. In 1943 he photographed the Manzanar Relocation Center, where Japanese Americans were interned by the U.S. government during World War II. In 1960 he was commissioned to photograph scenes from all nine campuses of the University of California. In his later years he received several honorary doctoral degrees and other awards, including one named in his honor by the Wilderness Society. During these years he traveled widely, giving lecturers and exhibiting his works.

Gordon A. Parker

FURTHER READING

Adams, Ansel. *Ansel Adams: Four Hundred Photographs.* Edited by Andrea G. Stillman. Boston: Little, Brown, 2007.

Alinder, Mary Street. *Ansel Adams: A Biography.* New York: Henry Holt, 1996.

Lichtenstein, Therese. *Master of Light: Ansel Adams and His Influences.* New York: New Line Books, 2006.

Amory, Cleveland

CATEGORIES: Activism and advocacy; animals and endangered species

IDENTIFICATION: American author and animal rights activist

BORN: September 2, 1917; Boston, Massachusetts

DIED: October 14, 1998; New York, New York

SIGNIFICANCE: Amory's decades of activism for animal rights and animal protection saved thousands of animals from extermination and helped bring the issue of cruelty to animals into the public spotlight.

As an author, Cleveland Amory is perhaps best known for his best-selling books featuring his cat, Polar Bear: *The Cat Who Came for Christmas* (1987), *The Cat and the Curmudgeon* (1990), and *The Best Cat Ever* (1993). He also published a number of social history studies, and in 1974 he published *Man Kind? Our Incredible War on Wildlife*, a work that has been credited with influencing the antihunting movement in the United States.

From the time he was a young child, Amory harbored a dream to create a sanctuary for animals where they would be protected from harm and allowed to roam free. To this end, he established the Fund for Animals, a nonprofit organization, in 1967. In 1977, the Fund for Animals initiated its first major rescue of animals when the U.S. Park Service scheduled the extermination, by shooting, of all the wild burros living in the Grand Canyon in Arizona. Over the next two years, Amory's organization orchestrated a helicopter airlift of 577 burros from the floor of the Grand Canyon, some seven thousand feet below its rim. An intensive, nationwide adoption campaign was conducted for the burros; those animals not adopted found homes at the Black Beauty Ranch, an animal sanctuary founded by Amory in Murchison, Texas, in 1979. Two years later, the Fund for Animals rescued another 5,000 burros that had been earmarked for destruction at Death Valley National Monument and the Naval Weapons Center at China Lake, both in California.

In the early 1980's the organization rescued more than sixty wild Spanish Andalusian goats from San Clemente Island off the coast of California after the U.S. Navy had decided to eradicate the island's population of the animals, saying that the move was mandated by the Endangered Species Act (1973) because the goats were eating the native, endangered species of vegetation on the island. The Navy intended to allow hunters to kill the animals for a fee. The Fund for Animals arranged a helicopter airlift of the animals to temporary quarters in San Diego, California, and another adoption campaign was successfully conducted.

Black Beauty Ranch, which was renamed Cleveland Amory Black Beauty Ranch in 2004 to honor its founder, has been home to thousands of domestic and exotic animals rescued from neglectful or abusive situations. Among the animals given sanctuary at the ranch since its founding have been chimpanzees, kangaroos, wild horses, buffalo, and elephants.

Through his work with the Fund for Animals, Amory developed the resources to conduct high-profile rescues of large numbers of animals, and the organization's continuing efforts have received much national and international publicity. This publicity has helped raise the public's consciousness about issues of cruelty and neglect toward animals.

Karen E. Kalumuck

FURTHER READING

Greenwald, Marilyn S. *Cleveland Amory: Media Curmudgeon and Animal Rights Crusader.* Hanover, N.H.: University Press of New England, 2009.

Marshall, Julie Hoffman. *Making Burros Fly: Cleveland Amory, Animal Rescue Pioneer.* Boulder, Colo.: Johnson Books, 2006.

Animal rights movement

CATEGORIES: Activism and advocacy; animals and endangered species; philosophy and ethics

IDENTIFICATION: Social movement involving groups and individuals concerned with the basic rights and welfare of animals

SIGNIFICANCE: The animal rights movement has brought about a number of changes in laws as well as in public perceptions of issues involving animals, but debates continue concerning the extent of the rights to which animals are entitled.

People involved in the animal rights movement share philosophical beliefs based on the idea that all animals are entitled to an equal claim on life and liberty and possess the same rights to existence as humans. Animal rightists oppose those who believe that animals exist for human use as objects of study and experimentation, as food, as beasts of burden, or as objects of amusement and recreation.

The philosophical concept of animal rights arose during the seventeenth and eighteenth centuries along with the development of biological science. The growing interest in biology gave rise to a sort of sideshow in which living, conscious dogs were cut open so that the animals' internal organs could be displayed to crowds of onlookers. A variety of blood sports were popular as well, including bullbaiting and bearbaiting. In these, a bull or a bear was chained in a ring along with one or more dogs that were trained to attack the larger animals. Dogfighting, in which various terrier breeds were encouraged to attack each other, was also popular.

BIRTH OF ANIMAL RIGHTS

In 1824 the Royal Society for the Prevention of Cruelty to Animals (RSPCA) was founded in Great Britain to enforce new anticruelty laws. However, the laws and their enforcement had little, if any, effect in rural areas, which were far from the watchful eyes of the police or RSPCA agents. On many farms, animals were still kept in filthy conditions and beaten if they balked at hauling overloaded wagons. The slaughter of animals for market was carried out as simply and quickly as possible.

Biomedical research, especially in human anatomy and physiology, advanced rapidly in Europe during the early to mid-nineteenth century. While anatomical study could be satisfied with human corpses, physiologists required living material, and animals became their targets. Although many of the animals used in medical research were rats and mice, dogs were also frequently used. Many of the animals were stolen pets, while others were strays that were found roaming on the streets. The treatment that dogs received in medical laboratories varied, but, for the purposes of good science, the animals had to be maintained and treated in clean, sanitary, and relatively stress-free environments. Many dogs died at the hands of medical researchers, and opposition to the practice quickly grew. Objections came from members of the general public, who had heard stories of both real and imagined horrors suffered by animals in the experiments.

The British government sought to quiet the complaints by passing the Cruelty to Animals Act of 1876. The act did not prohibit the practice of experimenting on live animals; rather, it set regulatory procedures that had to be followed in the laboratories. Animal rights were no less an issue in the United States at that time. The first documented humane society in the United States was the American Society for the Prevention of Cruelty to Animals (ASPCA), incorporated in 1866. Another pioneer group, the American Antivivisection Society (AAS), was founded in 1883.

The animal rights movement in the United States was relatively quiet until the first Earth Day in 1970, after which it rapidly expanded. Of the more than eighty animal rights and animal welfare organizations in the United States, fifty-seven (70 percent) were founded after 1970. In addition to the older groups, among the organizations are such diverse associations as Actors and Others for Animals, the Coalition for Non-Violent Food, the Animal Political Action Committee, and the American Fund for Alternatives to Animal Research. The list also includes several adversarial and confrontational groups, such as Greenpeace, People for the Ethical Treatment of Animals (PETA), and the Animal Liberation Front (ALF).

ANIMAL RIGHTS ISSUES

One issue that quickly attracted animal rights proponents was the plight of the whales. For centuries whaling was conducted from sailing ships with handheld harpoons in the manner made famous in Herman Melville's novel *Moby Dick* (1851). Even with such crude equipment and methodology, whalers reduced the whale population in the Atlantic Ocean and turned their attention to the Pacific Ocean. In the late nineteenth century, steam (and, later, diesel) vessels and cannon-fired harpoons increased the whalers' efficiency. The methods of the whalers aroused the ire of many people. Frequently, a harpooned whale was forced to tow the steel "catcher" ship for hours until the animal succumbed to the injuries from the explosive-headed harpoon. Another whaling technique was to harpoon and kill a whale calf. The mother and other adults hovered around the injured or killed calf and were in turn harpooned.

Economics rather than animal rights brought about the formation of the International Whaling Commission (IWC) in 1946. The commission was established to manage the whale stocks. It had no regulatory authority, however, and Norway, Iceland, and Japan continued to hunt whales despite the recommendations of the IWC. These nations have defended their activities as a sustainable use of a natural resource. Opponents view whaling as an archaic activity and a violation of animal rights. In 1972 the U.S. government passed the Marine Mammal Protection Act as a move to protect the whales. The apparent recovery of the Pacific gray whale stocks suggests that the act may have been a step toward achieving the goals of animal rights activists.

Dolphins are also at risk, but from

Major Animal Rights Organizations

Organization	Founded
American Society for the Prevention of Cruelty to Animals	1866
American Anti-Vivisection Society	1883
Animal Defense League	1934
Friends of Animals	1957
Fund for Animals	1967
Actors and Others for Animals	1971
American Fund for Alternatives to Animal Research	1977
People for the Ethical Treatment of Animals	1980
Alliance for Animals	1988

indirect human exploitation. In the eastern tropical Pacific Ocean, yellowfin tuna frequently swim below pods of dolphins, and commercial fishermen learned to set their nets around the schools of dolphins to capture the tuna. As the fish are netted, the trapped dolphins drown. The killing of as many as 132,000 dolphins each year in this manner led to protests by animal rights groups that included a boycott of canned tuna. In response to the public outcry, the U.S. government instituted regulations that require both domestic and foreign fishers to follow practices that release the dolphins from their nets and still retain most of the tuna. Within five years after the regulations were put in place, the accidental catch and kill of dolphins was reduced to 25 percent of what it had been. The reduced tuna catches, however, forced many U.S. fishers out of the industry, leaving a void that was quickly filled by foreign fishers.

The animal rights movement has also brought increased scrutiny to the fur trade and to circuses, zoos, theme parks, and any other activity in which live animals are used. For example, in February, 1995, when the Ringling Brothers and Barnum & Bailey Circus was preparing to visit Richmond, Virginia, the show's management asked the news media not to mention the time of the circus train's arrival, because the circus wanted to move its animals from the rail yard to the show grounds with as little fanfare as possible. The circus had received threats from animal rights groups and sought to avoid any confrontation and possible risks to animals and the public. Although the animal rights movement has brought about a number of changes in laws as well as in public perceptions of animal issues, it is unlikely that ongoing controversies surrounding animal rights will soon be resolved.

Albert C. Jensen

FURTHER READING

Beers, Diane L. *For the Prevention of Cruelty: The History and Legacy of Animal Rights Activism in the United States.* Athens: Swallow Press/Ohio University Press, 2006.
Fellenz, Marc R. *The Moral Menagerie: Philosophy and Animal Rights.* Urbana: University of Illinois Press, 2007.
Franklin, Julian H. *Animal Rights and Moral Philosophy.* New York: Columbia University Press, 2005.
Garner, Robert. *The Political Theory of Animal Rights.* New York: Manchester University Press, 2005.
Harnack, Andrew, ed. *Animal Rights: Opposing Viewpoints.* Farmington Hills, Mich.: Greenhaven Press, 1996.
Sunstein, Cass R., and Martha C. Nussbaum, eds. *Animal Rights: Current Debates and New Directions.* New York: Oxford University Press, 2006.
Whisker, James B. *The Right to Hunt.* 2d ed. Bellevue, Wash.: Merril Press, 1999.

Animal testing

CATEGORIES: Animals and endangered species; philosophy and ethics
DEFINITION: Use of nonhuman animals for research purposes, particularly medical research
SIGNIFICANCE: Animal testing is an integral component of modern science, product testing, and education. Most significant developments in medicine directly or indirectly rely on animal testing. Public debate about the moral and legal status of animals in society has resulted in numerous regulations on animal testing, yet it continues to be a controversial subject.

Animal testing, also known as animal experimentation, animal research, in vivo testing, and vivisection, is used to advance pure and applied research. Behavior, development, evolution, and genetics research are all forms of pure research involving animals. Applied research includes medical research, defense research, and toxicology studies for drugs, food additives, pesticides, and cosmetics. The use of animals in medical education and training is typically considered to be a form of animal testing as well.

RATIONALE, SCOPE, AND REGULATIONS

The underlying rationale for the use of animal testing is that living organisms provide interactive, dynamic systems that scientists can observe and manipulate in order to understand normal and pathological functioning as well as the effectiveness of medical interventions. The vast majority of animal testing is for human benefit and relies on the physiological and anatomical similarities between humans and other animals. The term "animal model" refers to the use of live animals to study particular biological processes with the end of extrapolating that information to other animals, particularly humans.

Many species are used in animal testing. Nematode

worms and fruit flies are commonly used invertebrates. Zebra fish and mice are commonly used vertebrates. While it is difficult to ascertain the exact number of animals used in research, it is estimated that between 50 million and 100 million vertebrates are used annually worldwide. Approximately 90 percent of the animals used in research are mice and rats.

A chief moral concern raised in regard to animal testing is the pain and suffering it involves. Sentience, or subjective awareness, particularly of pain and pleasure, is common to all vertebrates. Evidence for sentience in invertebrates is generally absent, however, and for this reason research on invertebrates is largely unregulated. Cephalopods (the class of animals that includes octopi and squid) are notable exceptions and are covered by regulations in several countries owing to evidence for their sentience.

A societal consensus exists that animal testing for the advancement of science and medicine is justified, provided that there are no alternatives, the use of animals is kept to a minimum, and that animal pain and distress is minimized. Supporters of animal testing commonly cite the number of major medical advances that have resulted from the practice.

Animal testing is heavily regulated in many countries. Regulations have changed significantly since the mid-twentieth century, and they differ in the numbers of species covered, the kinds of animal welfare protections offered, and the regulatory approaches taken. In the United States, animal testing is governed by two federal statutes: the Animal Welfare Act and Regulations of 1966 (AWAR) and the Health Research Extension Act of 1985, the provisions of which are carried out in the Public Health Service Policy on Humane Care and Use of Laboratory Animals (PHS Policy). AWAR establishes the minimum acceptable standards of care and treatment for certain animals in research, testing, experimentation, exhibition purposes, and use as pets. AWAR covers all warm-blooded animals yet specifically excludes birds, mice, and rats bred for research purposes, as well as animals used for food, fiber, or many forms of agricultural research. PHS Policy, which applies to all research funded by the National Institutes of Health, applies to all live vertebrates used for research purposes.

Animal Welfare Act

The main provisions of the Animal Welfare Act of 1966 follow:

An Act,

To authorize the Secretary of Agriculture to regulate the transportation, sale, and handling of dogs, cats, and certain other animals intended to be used for purposes of research or experimentation, and for other purposes.

Be it enacted by the Senate and House of Representatives of the United States of America in Congress assembled. That, in order to protect the owners of dogs and cats from theft of such pets, to prevent the sale or use of dogs and cats which have been stolen, and to insure that certain animals intended for use in research facilities are provided humane care and treatment, it is essential to regulate the transportation, purchase, sale, housing, care, handling, and treatment of such animals by persons or organizations engaged in using them for research or experimental purposes or in transporting, buying, or selling them for such use.

SEC. 2. When used in this Act— . . .

(d) The term "dog" means any live dog (*Canis familiaris*);
(e) The term "cat" means any live cat (*Felis catus*);
(f) The term "research facility" means any school, institution, organization, or person that uses or intends to use dogs or cats in research, tests, or experiments, and that (1) purchases or transports dogs or cats in commerce, or (2) receives funds under a grant, award, loan, or contract from a department, agency, or instrumentality of the United States for the purpose of carrying out research, tests, or experiments;
(g) The term "dealer" means any person who for compensation or profit delivers for transportation, or transports, except as a common carrier, buys, or sells dogs or cats in commerce for research purposes;
(h) The term "animal" means live dogs, cats, monkeys (nonhuman primate mammals), guinea pigs, hamsters, and rabbits.

Opposition and Alternatives

Opposition to animal testing is diverse. Disagreement with the practice is based on both scientific and ethical grounds, and it varies both according to species and according to purpose. Research using primates, monkeys, cats, and dogs is particularly controversial. Cosmetics testing on animals is controversial because many consider the benefit of yet another cosmetic product to be of dubious value when weighed against animals' interests. Although cosmetics testing on animals remains legal in the United States, it has been banned in several countries, including the United Kingdom, and the European Union began phasing it out in 2009. Some object to the use of animals in education, asserting that such use encourages the view that animals are objects to be manipulated rather than beings deserving of compassion or respect.

Those opposed to animal testing on scientific grounds cite the unreliability of predicting effects in humans based on animal models. Some argue that animal testing is not cost-effective; they assert that, given the substantial costs of conducting animal tests, which often last years and cost millions of dollars, the goal of improving human health would be more fully and efficiently realized through a reallocation of funding to implement existing medical technologies more widely. Some argue that much animal testing is immoral because the animal suffering caused is greater than the expected benefits to humans. The stronger animal rights view is that each animal has inherent moral worth, which prohibits humans from using them as experimental subjects for any reason.

First articulated by scientists William M. Russell and Rex L. Burch, the "three R's"—replacement, reduction, and refinement—are influential guiding principles for the humane use of animals in research. Replacement involves seeking to increase alternatives to animal testing that generate the desired research data without the use of sentient animals. Examples of replacement include the use of computer models, epidemiological data, tissue cultures, isolated organs, and nonsentient animals. Reduction is the effort to obtain comparable data using fewer animals or to obtain more data using the same number of animals. Refinement involves favoring research protocols that alleviate or minimize animal pain and distress through the use of analgesics, veterinary care, improved living quarters, and enrichment. Further development and increased implementation of alternatives to and refinement of animal testing is an area of common ground between animal advocates and animal researchers.

Joel P. MacClellan

FURTHER READING

Carbone, Larry. *What Animals Want: Expertise and Advocacy in Laboratory Animal Welfare Policy.* New York: Oxford University Press, 2004.

Cothran, Helen, ed. *Animal Experimentation: Opposing Viewpoints.* San Diego, Calif.: Greenhaven Press, 2002.
Guerrini, Anita. *Experimenting with Humans and Animals: From Galen to Animal Rights.* Baltimore: The Johns Hopkins University Press, 2003.
Monamy, Vaughan. *Animal Experimentation: A Guide to the Issues.* 2d ed. New York: Cambridge University Press, 2009.
Paul, Ellen Frankel, and Jeffrey Paul, eds. *Why Animal Experimentation Matters: The Use of Animals in Medical Research.* New Brunswick, N.J.: Transaction, 2001.

Antienvironmentalism

CATEGORIES: Philosophy and ethics; activism and advocacy
DEFINITION: Philosophy that holds that human beings' immediate economic and lifestyle needs are more important than concerns about the fate of other species and the general environment
SIGNIFICANCE: The long-standing debates that continue between antienvironmentalists and environmentalists have important influence on both legislators and policy makers as they address environment-related issues.

In the early twentieth century, environmentalism in the United States was largely fostered by wealthy sportsmen who saw the need to protect the outdoors in order to maintain satisfactory areas for their pursuits of hunting, fishing, and camping. The movement got a populist boost in 1962 from the publication of Rachel Carson's *Silent Spring*, which presents an easily understood account of the dangers of toxic substances in the environment. For the first time, the American public began to demand that laws be enacted to protect the environment and clean up land, water, and air that had already been polluted.

GROWTH OF THE ENVIRONMENTAL MOVEMENT

For several years the environmental movement gathered strength as the public voted into office politicians with environmental orientations. Public outcry surged against polluting companies, leading to boycotts of products. Grassroots, citizen-led efforts such as recycling programs and litter patrols gained support as the public became more educated and concerned about environmental issues. Among the issues that pitted environmentalists against the government and industry were toxic waste incineration, habitat destruction by logging and mining companies, and use of public lands, including national parks.

Two oil crises during the 1970's served to focus awareness on energy conservation and the need to develop alternatives to energy derived from fossil fuels. Many feared that oil supplies were dwindling, while others wished to end U.S. reliance on oil-exporting nations in the Middle East. One important result was a general reduction in the size of motor vehicles. This, along with other technological advances, helped lead to the development of cars that were more fuel-efficient. The research required to accomplish these changes, however, was very costly to automakers.

The 1970's were also characterized by landmark legislation that imposed strict limits on pollution output and resource use and also provided for the remediation of polluted land and water. Large fines were imposed for violations of the new laws, which were enforced by the newly formed U.S. Environmental Protection Agency (EPA). One of the most important and pivotal developments was the passage of the Comprehensive Environmental Response, Compensation, and Liability Act, or Superfund, which provided vast sums of public money for the cleanup of designated industrial and military waste dumps and other degraded sites. Signed into law by the U.S. Congress in 1980, Superfund's provisions allowed the government to bring lawsuits against the responsible parties, requiring them to help pay cleanup costs. In order to avoid fines, many industries were forced to develop and implement costly waste-processing technologies.

The political and economic situation began to change in the late 1970's as industry mounted a counteroffensive against environmental laws. Businesses contended with the burgeoning number of environmental regulations by finding and exploiting loopholes in legislation. A growing number of industries used stalling tactics and countersuits to delay or eliminate the need to implement required changes. Meanwhile, in the western United States, a coalition of loggers, miners, cattle ranchers, farmers, and developers demanded that the federal government transfer control of large tracts of federally owned land to individual states. Members of the so-called Sagebrush Rebellion felt that state ownership would give them more power to exploit the natural resources on the land.

Antienvironmentalist Backlash

A severe backlash against environmentalism began to occur when Ronald Reagan replaced Jimmy Carter as president of the United States in early 1981. Many environmental laws and regulations, which the new presidential administration viewed as barriers to economic progress, were weakened or abolished. A large number of federal judges who had started their careers during the 1960's retired, and they were replaced by politically conservative judges who began to interpret existing laws in favor of industry. The office of the EPA was weakened, and funding for environmental enforcement and remediation was slashed. Secretary of the Interior James Watt, who had been a leader of the Sagebrush Rebellion, promoted legislation to open previously protected areas to mining and oil exploration. The general public, experiencing growth and prosperity for the first time in many years, began to favor short-term economic gains and turned a blind eye to news of the weakening environmental movement.

The late 1980's saw the birth of the wise-use movement, which appeals to the pragmatic and optimistic aspects of human nature by asserting that some optimal balance of resource use and restoration is practicable and that technology, given time and funding, will develop workable solutions to existing environmental problems. This position assumes that human beings can understand the complex ecosystems involved well enough to know what these balances should be. Advocates of wise use believe that all public lands, including national parks, should be opened to mining and drilling. Like the Sagebrush Rebels, they also promote the strengthening of the rights of states and property owners to exploit resources with minimal federal regulation.

According to the tenets of wise use, the harvesting of timber from ancient forests would be followed by the planting of an equivalent acreage of saplings. Logging would be timed according to growth rates, and technology would produce fast-growing varieties of trees that would furnish adequate ecosystems for wildlife in the new forests. Environmentalists, in contrast, argue that ancient forests represent complex, irreplaceable ecosystems that cannot be substituted with new forests planted by logging companies. Similar disagreement exists regarding coastal wetlands, which provide vital habitat to numerous species and contain a high degree of biodiversity. Wetlands are frequently located in areas that are desired by real estate

> **Equating Environmentalism with Marxism**
>
> *Llewellyn H. Rockwell begins "An Anti-Environmental Manifesto" by comparing environmentalism with Marxism-Leninism:*
>
> The last Stalinist, Alexander Cockburn, has gone from attacking Gorbachev (for selling out Brezhnev) to defending Mother Earth. His new book, *The Fate of the Forest*, is both statist and pantheist.
>
> Cockburn, a man who supposedly cares about peasants and workers, instead decries their cutting down the Brazilian rainforests to farm and ranch. People are supposed to live in indentured mildewtude so no tree is touched.
>
> But Cockburn is part of a trend. All over Europe and the U.S., Marxists are joining the environmental movement. And no wonder: environmentalism is also a coercive utopianism—one as impossible to achieve as socialism, and just as destructive in the attempt. . . .
>
> Today we face an ideology every bit as pitiless and messianic as Marxism. And like socialism a hundred years ago, it holds the moral high ground. Not as the brotherhood of man, since we live in post-Christian times, but as the brotherhood of bugs. Like socialism, environmentalism combines an atheistic religion with virulent statism. But it ups the ante. Marxism at least professed a concern with human beings; environmentalism harks back to a godless, manless, and mindless Garden of Eden.
>
> If these people were merely wacky cultists, who bought acres of wilderness and lived on it as primitives, we would not be threatened. But they seek to use the state, and even a world state, to achieve their vision.
>
> And like Marx and Lenin, they are heirs to Jean Jacques Rousseau. His paeans to statism, egalitarianism, and totalitarian democracy have shaped the Left for 200 years, and as a nature worshipper and exalter of the primitive, he was also the father of environmentalism.
>
> During the Reign of Terror, Rousseauians constituted what Isabel Paterson called "humanitarians with the guillotine." We face something worse: plantitarians with the pistol.

developers wishing to build vacation homes or resorts. Environmental protection laws based on the tenets of wise use mandate that destroyed wetlands must be replaced by new wetlands of equivalent size. Again, environmentalists worry that too many threatened species would be lost in the process of destroying and replacing wetland areas.

SCIENTIFIC CONTROVERSY

Another significant trend in antienvironmentalism toward the end of the twentieth century involved public confusion over scientific debates concerning such topics as the ozone hole and global warming. Industry and government scientists often questioned and condemned dire predictions advanced by other scientists. The government frequently responded by requesting additional research before requiring vast, expensive reductions of known pollutants. Rather than believe the frightening scenarios painted by some scientists, many people sided with scientists who questioned the validity of these and other threats to the global environment.

Mainstream environmental organizations that had evolved from small groups of fervent individuals were now led by full-time professional lobbyists based in Washington, D.C. Details of new legislation were negotiated among industry, government, and environmental leaders. National environmental organizations grew increasingly cumbersome and expensive to run. Many began accepting large donations from the same industries they were trying to monitor, which created serious conflicts of interest. Top industry executives became members of the boards of directors for environmental organizations. At the same time, these corporations also made large donations to elected officials and thus gained access and influence in government. As the 1990's progressed, membership in large environmental groups began to decline as "donor fatigue" set in, questions persisted about the true urgency of various environmental issues, and cynicism arose about the possibility of environmental progress under such circumstances.

A relaxation of concern about environmental problems came about in the late 1990's in the wake of encouraging news about improvements of environmental indicators such as air-pollution levels of certain gases in the aftermath of implementation of cleaner energy production. For example, atmospheric levels of sulfur dioxide, which leads to acid rain, decreased in the United States and Europe after cleaner coal- and oil-burning technologies were implemented. The air-quality goals of many cities were met through a combination of fuel efficiency and "scrubber" smokestacks.

A controversial strategy advanced by a coalition of industry, government, and environmental leaders involves tradable pollution permits. According to the plan, the government assigns utilities a certain number of pollution units per year. An especially clean-running plant will not need all of its units and will be able to sell them to plants that exceed their allotments. This system has been criticized by many environmentalists, who contend that some utilities are able to buy their way out of the need to reduce pollution. The position thought to be antienvironmentalist in this context would maintain that the plan is a realistic method of controlling overall levels of pollution without putting older utilities out of business while they endeavor to upgrade their performance.

The revelation that environmental degradation can, in certain cases, be reversed over a fairly short span of time led to the argument on the part of antienvironmentalists that nature is surprisingly resilient, and therefore environmental protection does not need to be so stringent, costly, and regressive. Environmentalists counter such arguments with a call to remain vigilant and to include the health of the environment in national and global visions of the future.

Wendy Halpin Hallows

FURTHER READING

Dowie, Mark. *Losing Ground: American Environmentalism at the Close of the Twentieth Century.* Cambridge, Mass.: MIT Press, 1995.

Easterbrook, Gregg. *A Moment on the Earth: The Coming Age of Environmental Optimism.* New York: Penguin Books, 1995.

Helvarg, David. *The War Against the Greens: The "Wise-Use" Movement, the New Right, and the Browning of America.* Rev. ed. Boulder, Colo.: Johnson Books, 2004.

Hirt, Paul W. *A Conspiracy of Optimism: Management of the National Forests Since World War Two.* Lincoln: University of Nebraska Press, 1994.

Jacques, Peter J. *Environmental Skepticism: Ecology, Power, and Public Life.* Burlington, Vt.: Ashgate, 2009.

McKibben, Bill. *The End of Nature.* 1989. Reprint. New York: Random House, 2006.

Shabecoff, Philip. *A Fierce Green Fire: The American Environmental Movement.* Rev. ed. Washington, D.C.: Island Press, 2003.

Young, John. *Sustaining the Earth: The Story of the Environmental Movement—Its Past Efforts and Future Challenges.* Cambridge, Mass.: Harvard University Press, 1990.

Antinuclear movement

CATEGORIES: Activism and advocacy; nuclear power and radiation

IDENTIFICATION: Social movement comprising a loose collection of organizations and individuals opposed to nuclear weapons and nuclear power

SIGNIFICANCE: The antinuclear movement, which emerged during the 1950's and gained momentum during the 1960's, initially focused its attention on the proliferation of nuclear weapons and the threat of nuclear warfare. It later broadened its scope to include opposition to nuclear power plants and nuclear waste facilities. Antinuclear activists have had some success in raising public awareness of the dangers of nuclear proliferation and in using safety regulations and environmental law to slow the development of nuclear power, particularly in the United States.

Antinuclear activists generally believe that if nuclear weapons are available, these weapons will eventually be used; they therefore seek to rid the earth of all such armaments. By organizing seminars, rallies, protests, and public outreach campaigns, activists seek to stimulate public debate on issues previously left to insiders, and they help shape the political climate. For example, activists from the environmental organization Greenpeace drew international attention to French resumption of nuclear weapons testing in 1995. The outrage that this publicity fueled may have contributed to the French government's decision to conduct fewer tests than planned.

Antinuclear activists have often been successful in capturing favorable media coverage and mobilizing hundreds of thousands of citizens to protests. However, translating this success into practical results has often proved elusive. At a conference on abolishing nuclear weapons held at Boston College in October, 1997, American Friends Service Committee staffer David McCauley offered a possible explanation. He suggested that the movement is fundamentally anarchistic and that distrust of people in power stifles activists' ability to cooperate with established political practices.

NUCLEAR FREEZE AND DISARMAMENT GOALS

In the multilateral Nuclear Non-Proliferation Treaty (NNPT), which entered into force in 1970, five recognized nuclear weapons states—the United States, the United Kingdom, France, China, and the Soviet Union—agreed not to transfer nuclear weapons to states without such weapons. With the treaty, signatory nations committed to work toward nuclear disarmament, yet the arms race continued. Antinuclear activists claimed that arms negotiators were allowing themselves to be defeated by complexities and suggested a mutual and verifiable freeze by the United States and the Soviet Union on the testing, production, and deployment of all nuclear weapons. The campaign attracted sympathetic media coverage and gained momentum, so that by the late 1970's hundreds of thousands of people were attending massive rallies in the United States and Europe to protest the planned deployment of the ground-launched cruise missile (GLCM) and the Pershing II missile.

Leaders of the North Atlantic Treaty Organization (NATO), however, dared not ignore the complexities inherent in nuclear negotiations. In 1976 the Soviet Union began updating its forces by deploying SS-20 missiles. NATO viewed this as a destabilization of the balance of power because the SS-20's were mobile, more accurate, equipped with three warheads each, and able to fly 1,500 to 3,000 kilometers (930 to 1,860 miles) farther than the missiles they replaced. After all attempts at diplomatic negotiations with the Soviets failed, NATO responded in 1983 by deploying Pershing IIs and GLCMs, which were more accurate than the SS-20's but had less than one-half their range.

NATO and the Soviet Union eventually agreed to intrusive, on-site verification procedures. In late 1987, the two nations signed the Intermediate-Range Nuclear Forces Treaty, which completely eliminated an entire class of weapons: SS-20's, Pershing IIs, GLCMs, and all other ground-launched nuclear missiles with ranges between 500 and 5,500 kilometers (310 and 3,420 miles), along with their launchers, support facilities, and bases. The treaty also banned flight testing and production of these missiles. The antinuclear movement deserves credit for helping educate the public and adding to the pressure that pushed politicians to the bargaining table. However, based on history, it seems unlikely that the treaty would have come about had NATO not deployed its own weapons.

In 1995, the year the NNPT was extended indefinitely, an antinuclear initiative called Abolition 2000 was proposed. Within a few years, more than twelve hundred nongovernmental organizations on six continents had voiced support for it; by 2010 it had grown to a network of more than two thousand organizations in more than ninety countries. Abolition 2000 called for movement toward "clean, safe, renewable forms of energy production that do not provide the materials for weapons of mass destruction and do not poison the environment for thousands of centuries." It also called for negotiations on a nuclear weapons abolition convention that would establish a timetable for phasing out all nuclear weapons and include provisions for effective verification and enforcement.

At its 2003 annual meeting, Abolition 2000 launched a collaborative effort with Mayors for Peace, an international program established in 1982 in which the heads of city governments around the globe work to encourage city-by-city support of the abolition of nuclear weapons. By August 6, 2010 (the sixty-fifth anniversary of the first use of the atomic bomb), 4,069 cities in 144 countries and regions around the world had become members of the Conference of Mayors for Peace and were calling for complete nuclear disarmament by 2020.

Nuclear Power in the United States

Many members of the antinuclear movement are also against the development and expansion of nuclear power generation. They believe that nuclear technology is too dangerous and see reactor accidents caused by natural disasters, equipment failures, or human errors as inevitable. They do not believe that radioactive waste can be safely disposed of, and they see such waste as an unfair burden to pass on to future generations. Furthermore, many argue that terrorists or nations without their own nuclear arsenals could divert reactor materials to make nuclear weapons.

During the 1960's the antinuclear movement became concerned with the possible effects of low-level radiation from nuclear power plants. Scientists are divided as to whether low levels of radiation can cause increased incidence of cancer in communities located near nuclear power plants; whereas some studies point to significant risk, others report considerable evidence that the human body is able to repair damage caused by sufficiently low levels of radiation.

The antinuclear movement has been somewhat successful in using safety regulations and environmental law to effect change. By lobbying for stricter regulations and suing to force nuclear plants to follow safety regulations that they previously had been allowed to bypass, activists have contributed to making nuclear plants safer. These actions have also added to the costs of nuclear power plants and to delays in plant licensing. Using the courts to delay construction of nuclear plants has been a powerful tool for the movement. In 1967 construction time for a nuclear power plant in the United States averaged 5.5 years; by 1980 it had reached 12 years.

The price of electricity from nuclear plants has increased over time with lengthening construction time, the addition of safety features, and the tendency of nuclear power companies to try new designs instead of settling on a standard design. In 1976 the price of electricity from coal-fired and nuclear plants was nearly the same, but by 1990 nuclear power was twice as expensive as coal power in the United States. One goal of the antinuclear movement was reached in the United States: With economics against them, planners ceased construction of new nuclear power plants.

The U.S. Nuclear Regulatory Commission (NRC) eventually combined the construction and operating licensing procedures to minimize delays. Also, the industry significantly reduced operating costs. These factors, combined with mounting concerns about the finite nature of fossil-fuel supplies and the potential of the burning of such fuels to affect global climate, made nuclear power a more attractive prospect by the early twenty-first century. Some previously antinuclear environmentalists reconsidered their stance on nuclear power and came out in favor of it as an alternative to fuels that produce greenhouse gas emissions.

Nuclear Power in Asia and Europe

According to the International Atomic Energy Agency, in 2008 nuclear power provided 19.7 percent of the electricity in the United States, 24.9 percent in Japan, 28.8 percent in Germany, 42.0 percent in Sweden, and 76.2 percent in France. Countries that must import much of their fuel, such as Japan and France, find nuclear power particularly attractive.

Antinuclear sentiment led Sweden to announce a phaseout of nuclear power beginning in 1998. The shutdown was delayed, however, for several reasons, including lack of replacement power, fears by workers and industry that the shutdown would lead to higher energy prices and exacerbate unemployment, and a lawsuit by a nuclear plant's owners seeking indemnification. In 2009, citing climate concerns, the Swedish government officially abolished its phaseout scheme. Some 52 percent of the Swedes who responded to a 2010 survey favored keeping nuclear power and replacing old reactors with new ones.

As part of the price of forming a ruling coalition with Germany's Social Democratic Party, the environmentalist Green Party extracted a promise that Germany's nuclear power reactors would eventually be eliminated. Shutdown plans were made official in 2002. Germany's government faced problems similar to Sweden's in trying to phase out nuclear power and felt additional pronuclear pressure from England and France, which held contracts worth $6.5 billion to reprocess German nuclear fuel. By 2008 concerns about power availability, costs, and carbon dioxide emission reduction goals had prompted Germany to explore how to keep its existing nuclear power plants online, even as thousands of protesters turned out in opposition to the shipment of nuclear reactor waste from France to a German storage site in Gorleben.

Scientific Groups

A few scientific groups have been especially influential in the area of nuclear arms control. Many scientists who developed the atomic bomb were against leaving decisions about its uses to a few elite government and military officials. Immediately after World War II, some of them formed the Atomic Scientists of Chicago and began publication of the *Bulletin of the Atomic Scientists*. Another group formed the Federation of Atomic Scientists. The efforts of these scientists contributed to the founding in 1946 of the Atomic Energy Commission, a civilian agency that took control of the materials, facilities, production, research, and information relating to nuclear fission from the military.

In response to the escalating arms race, in 1955 the physicist Albert Einstein and the philosopher, mathematician, and social critic Bertrand Russell published a manifesto in which they called upon scientists to assemble and appraise the perils of nuclear weapons. Cyrus Eaton, a wealthy industrialist and admirer of Russell, invited twenty-two scientists from both sides of the Iron Curtain to a conference in July, 1957, that was held in Eaton's summer home in the small village of Pugwash, Nova Scotia. The scientists were able to function as icebreakers between governments; in fact, some were government advisers. More Pugwash conferences followed, providing invaluable contacts, networks, and facts for those involved in arms control. The antinuclear movement in general and scientific organizations in particular were instrumental in bringing about several treaties, including the Limited Test Ban Treaty of 1963, in which signatory nations agreed to end aboveground nuclear testing.

The Union of Concerned Scientists (UCS) was formed in 1969 by faculty and students at the Massachusetts Institute of Technology who felt that too much emphasis was being placed on research with military applications and not enough on research that could address environmental and social concerns. The UCS subsequently grew to become a coalition of scientists and citizens across the United States and expanded its focus to include renewable energy and other environmental issues. The UCS combated the establishment of an antiballistic missile defense system in the United States and played a key role in defeating a scheme to rotate two hundred MX missiles among 4,600 protective silos. The plan would have cost $37 billion and would have swallowed vast tracts of the western desert of the United States.

The UCS has also waged campaigns calling for U.S. support of the 1996 Comprehensive Nuclear-Test-Ban Treaty, which prohibits all nuclear explosions (the United States signed the treaty in 1996 but has not ratified it). The UCS has worked in support of other nuclear disarmament treaties as well, including a series of agreements between the United States and the Soviet Union (later the Russian Federation) limiting the number of nuclear warheads and delivery vehicles that these nations may deploy: the 1991 Strategic Arms Reduction Treaty (START), the 2002 Strategic Offensive Reductions Treaty (SORT), and the 2010 New START.

Alternative Energy Sources

One of the great challenges facing the antinuclear movement is that of finding alternative sources of energy. Many nonnuclear power plants have serious environmental effects of their own. By some estimates, cardiopulmonary illness linked to air pollution accounts for tens of thousands of deaths in the United States each year. Power plants are major contributors to such pollution, with older, coal-fired plants being among the worst offenders. Burning coal not only produces copious amounts of carbon dioxide but also releases particulates, sulfur compounds, lead, arsenic, mercury, naturally occurring radioactive elements, and other harmful elements. Pollution-control equipment increases the construction and operating costs of new coal-fired plants. Further, running this equipment expends energy, and the equipment can only reduce pollution, not eliminate it.

Many segments of the antinuclear movement actively support the development of alternative energy sources, such as solar, geothermal, wind, and biomass power. The United States leads the world in the use of alternative energy sources, but by 2008 the U.S. Department of Energy reported that renewable energy sources (including hydropower) accounted for only 9 percent of the electricity generated in the nation.

Charles W. Rogers
Updated by Karen N. Kähler

Further Reading

Bodansky, David. *Nuclear Energy: Principles, Practices, and Prospects.* 2d ed. New York: Springer, 2004.

Cooke, Stephanie. *In Mortal Hands: A Cautionary History of the Nuclear Age.* New York: Bloomsbury, 2009.

Cortright, David, and Raimo Väyrynen. *Towards Nuclear Zero.* New York: Routledge, 2010.

Evangelista, Matthew. *Unarmed Forces: The Transnational Movement to End the Cold War.* 1999. Reprint. Ithaca, N.Y.: Cornell University Press, 2002.

Giugni, Marco. *Social Protest and Policy Change: Ecology, Antinuclear, and Peace Movements in Comparative Perspective.* Lanham, Md.: Rowman & Littlefield, 2004.

Gusterson, Hugh. *Nuclear Rites: A Weapons Laboratory at the End of the Cold War.* Berkeley: University of California Press, 1996.

Peterson, Christian. *Ronald Reagan and Antinuclear Movements in the United States and Western Europe, 1981-1987.* Lewiston, N.Y.: Edwin Mellen Press, 2003.

Price, Jerome. *The Antinuclear Movement.* Rev. ed. Boston: Twayne Publishers, 1990.

Wittner, Lawrence S. *Confronting the Bomb: A Short History of the World Nuclear Disarmament Movement.* Stanford, Calif.: Stanford University Press, 2009.

_____. *Toward Nuclear Abolition: A History of the World Nuclear Disarmament Movement, 1971 to the Present.* Stanford, Calif.: Stanford University Press, 2003.

Audubon, John James

Categories: Activism and advocacy; animals and endangered species

Identification: French American naturalist and wildlife artist

BORN: April 26, 1785; Les Cayes, Saint-Domingue (now Haiti)
DIED: January 27, 1851; New York, New York
SIGNIFICANCE: Through his unique paintings and his writings, Audubon demonstrated ecological relationships among organisms and set new standards for field observation. By illustrating the beauty of birds and animals, he helped to lay the foundation for a national environmental consciousness in the United States.

John James Audubon was born Jean Jacques Audubon, the son of a French naval officer, on April 26, 1785. He grew up in Saint-Domingue (now Haiti) and Nantes, France. At the age of eighteen he was sent to live on a farm at Mill Grove, near Philadelphia, Pennsylvania, to escape induction into Napoleon Bonaparte's army. Audubon spent his days in Pennsylvania roaming the woods and observing, collecting, and sketching wildlife. He experimented with bird banding and developed techniques to mount bird specimens so that he could draw them in lifelike poses. During this time he began thinking of himself as an American and began to call himself John James.

As Audubon matured, the hobby of his youth became an obsession. The time he spent outdoors teaching himself to live in the forest and understand its inhabitants cost Audubon dearly in financial terms and tried the patience of his wife, Lucy. However, it gave him unparalleled insights into the lives of his subjects. A perfectionist who periodically tore up drawings he considered less than his best, Audubon developed a distinctive style of illustration, creating drawings and paintings portraying birds and animals in lively interaction with their surroundings.

Audubon established a frontier store and a sawmill in Henderson, Kentucky, but these ventures failed because he so often neglected his business to pursue his art. Forced to declare bankruptcy in 1819, Audubon first thought of giving up painting altogether, then decided to devote himself completely to it. He resolved to publish a portfolio of American birds that would be more complete and more accurately illustrated than any previous work. Lucy taught school to support the family while he worked furiously to produce the necessary sketches and find a publisher. When the work appeared, Audubon suffered mixed reviews from the American scientific and artistic establishments but found a receptive audience in Europe.

In 1826 Audubon made arrangements with an engraver to make folio-sized prints of his illustrations and publish them under the title *The Birds of America*. This large-scale book would present each bird study as a life-size portrait and preserve the fine details of Audubon's watercolors. The hand-colored printing process was laborious and very expensive. To finance the project, Audubon sold subscriptions to a series of separate folios, each containing five engravings. It took him twelve years to complete the 435 plates. Only 176 subscriptions were sold, but this was enough to cover Audubon's expenses. (In the twenty-first century, first-edition copies of *The Birds of America* are prized by collectors, commanding high prices when offered for sale. In December, 2010, a complete first-edition copy was sold at auction for approximately $11.5 million, a record price for a printed book.)

With the appearance of *The Birds of America* in 1838

Naturalist and wildlife artist John James Audubon. (Library of Congress)

Audubon established an international reputation. The smaller and less expensive edition of the book, which included excerpts from his *Ornithological Biography* (1839), was a popular and financial success. Finally, Audubon was able to buy an estate on the Hudson River in New York, where he and Lucy mentored many young scholars, artists, and naturalists. One of these students, George Bird Grinnell, founded the first Audubon Society, an organization devoted to promoting bird study and conservation, in 1886. (The present-day National Audubon Society was incorporated in 1905.) As Audubon's energy and eyesight waned, his sons helped him complete paintings for *The Viviparous Quadrupeds of North America* (1849-1854).

With his unique paintings, Audubon demonstrated ecological relationships among organisms by illustrating their food plants, nesting sites, competition, and predators. In his writings, Audubon set new standards for field observation and foresaw the threat of species extinction. Above all, by illustrating the beauty of birds and animals, he promoted the popular study of natural history, helping to lay the foundation for a national environmental consciousness in the United States.

Robert W. Kingsolver

FURTHER READING

Rhodes, Richard. *John James Audubon: The Making of an American.* New York: Alfred A. Knopf, 2004.

Souder, William. *Under a Wild Sky: John James Audubon and the Making of "The Birds of America."* New York: North Point Press, 2004.

Back-to-the-land movement

CATEGORIES: Philosophy and ethics; activism and advocacy
IDENTIFICATION: Social movement based on the values of self-sufficiency and human harmony with nature
DATES: 1960's-1970's
SIGNIFICANCE: The back-to-the-land movement exemplified a practice that sought to give greater meaning to everyday life through adherence to values of self-sufficiency, simplicity, freedom, and, most important, anticonsumerism.

The expression "back to the land" is commonly used in reference to a North American social and countercultural phenomenon that started during the mid-1960's and continued well into the 1970's. The historical roots of the movement can be traced to Thomas Jefferson's agrarian vision and to the practice of self-reliance espoused by nineteenth century philosophers and essayists Ralph Waldo Emerson and Henry David Thoreau. Those who took part in the so-called back-to-the-land movement migrated from cities to rural areas because they had become increasingly disenchanted with the growing urban and industrial culture; they were attracted to a simple sort of daily life based on a set of values and choices that they saw as being in tune with an agrarian way of life and thus in greater harmony with the natural world.

Among the activities undertaken by those who joined the movement were building homes with natural materials; setting up systems to generate alternative forms of energy, such as solar and wind power; and growing their own food. They also faced choices regarding what types of livelihoods they should pursue and whether to work at home or outside the home; in addition, they somehow had to reconcile their desire to give up the rampant consumerism in society with their need to make a living. Members of the movement were arguably more interested in "making a life" for themselves than they were in "making a living," and many chose to live very simply, at the mercy of nature, and endure some of the inconveniences this may occasionally entail rather than indulge in what they viewed as the rampant and alienating consumerism of North American society at large.

By the end of the 1970's the movement evolved into one that took up the causes of the growing environmental movement, as people became more concerned with sustainable and holistic living. It has been estimated that by this time well over 1 million people had moved from the cities to rural areas. In the 1980's, however, a significant decline was noted in the number of people interested in leaving consumer culture for a simpler life in the countryside. This is explained in part by the booming and widespread prosperity of that decade. According to some observers of these trends, the 1990's saw a return to more environmentally conscious lifestyles among many North Americans, particularly in view of increasing awareness of global ecological crises.

Nader N. Chokr

FURTHER READING

Agnew, Eleanor. *Back from the Land: How Young Americans Went to Nature in the 1970's, and Why They Came Back.* Chicago: Ivan R. Dee, 2004.

Jacob, Jeffrey C. *New Pioneers: The Back-to-the-Land Movement and the Search for a Sustainable Future.* 1997. Reprint. University Park: Pennsylvania State University Press, 2006.

Nearing, Helen, and Scott Nearing. *Living the Good Life: How to Live Sanely and Simply in a Troubled World.* 1954. Reprint. New York: Schocken Books, 1987.

Berry, Wendell

CATEGORIES: Activism and advocacy; agriculture and food
IDENTIFICATION: American author of books on conservation and agrarianism
BORN: August 5, 1934; Henry County, Kentucky
SIGNIFICANCE: Berry's integrated professions of farmer, writer, and critic of industrial development have placed him among the major figures of the twentieth century in both conservation and literature.

Born to a tobacco farm family during the Great Depression, Wendell Berry grew up in a simple environment of small farms that practiced crop diversification, organic fertilizing, and use of draft animals. Berry's family had deep roots in the community, as did their neighbors. Farmers in Henry County, Kentucky, were largely self-sufficient, depending little on resources beyond their region.

Berry's rural upbringing affected every facet of his adult life. After receiving bachelor's and master's degrees in English from the University of Kentucky at Lexington, he was awarded a prestigious Wallace Stegner Fellowship in creative writing at Stanford University in 1958. This opportunity moved Berry into a circle of scholars and writers, notably Stegner, who at that time was a prominent novelist and conservationist. In 1960 Berry returned to Henry County. He spent a brief time in France and Italy on a Guggenheim Fellowship and later held teaching positions at New York University, Georgetown College, the University of Cincinnati, Bucknell University in Pennsylvania, and Stanford University. He has lived most of his life, however, with his wife Tanya Amyx Berry (they married in 1957) in his native north central Kentucky, where the couple farms. He was an English professor at the University of Kentucky from 1964 to 1977 and from 1987 to 1993.

Berry is an award-winning writer of more than forty books of poetry, fiction, and essays. He is considered by many to be one of the most important nature poets of his generation. He gained widespread recognition for his early volumes *The Broken Ground* (1964) and *Openings* (1968), both of which center on rural themes that he would continue to explore in later works such as *Given* (2005), *The Mad Farmer Poems* (2008), and *Leavings* (2009). Berry's writing on environmental issues focuses mainly on agriculture and the simple life. In his most successful novel, *The Memory of Old Jack* (1974), a ninety-two-year-old farmer relives his simple, agrarian life in flashbacks. Old Jack reflects the longing of Berry to return to his pre-World War II life of rural self-sufficiency. Among Berry's best-known nonfiction works are *The Long-Legged House* (1969), *The Hidden Wound* (1970), *A Con-

Author Wendell Berry. (©Dan Carraco/Courtesy, North Point Press)

tinuous Harmony (1972), *The Unsettling of America: Culture and Agriculture* (1977), and *The Gift of Good Land* (1981). In later essays such as those found in *Citizenship Papers* (2003), he examines the environmental, sociological, and political consequences of global industrialism.

Berry's writings emphasize the connectedness of human beings with the rest of nature. He is critical of the destruction of the land by mechanized monoculture farming, use of pesticides and fertilizers, clear-cutting of forests, and strip mining. He decries the movement away from the family farm to corporate farming, asserting that the corporate sector has killed rural America.

Berry views farming as an art and prioritizes ecology over economics, which values efficiency and specialization as means of maximizing income in the short term. In his view, technology has dehumanized agriculture by replacing the self-fulfilling labor of farmers and their families. In addition, technology has driven people from their land into the cities, generating social and environmental problems in urban areas.

Berry believes that a return to sustainable agriculture is an ecological imperative for maintaining a high quality of life. In his writings he praises the Amish for their stewardship of the land and their farming on a scale appropriate to the needs of their communities. He likewise admonishes Christians to heed the biblical message that "the earth is the Lord's and the fullness thereof."

Ruth Bamberger
Updated by Karen N. Kähler

Further Reading

Berry, Wendell. *The Art of the Commonplace: The Agrarian Essays of Wendell Berry*. Edited by Norman Wirzba. Washington, D.C.: Counterpoint, 2002.

Bonzo, J. Matthew, and Michael R. Stevens. *Wendell Berry and the Cultivation of Life: A Reader's Guide*. Grand Rapids, Mich.: Brazos Press, 2008.

Peters, Jason. *Wendell Berry: Life and Work*. Lexington: University Press of Kentucky, 2007.

Smith, Kimberly K. *Wendell Berry and the Agrarian Tradition: A Common Grace*. Lawrence: University Press of Kansas, 2003.

Bookchin, Murray

CATEGORIES: Activism and advocacy; urban environments
IDENTIFICATION: American ecological activist, author, and anarchist thinker
BORN: January 14, 1921; New York, New York
DIED: July 30, 2006; Burlington, Vermont
SIGNIFICANCE: Bookchin, the creator of the concept of social ecology, suggested in the 1960's that the prosperity of the post-World War II United States had been bought at the price of serious harm to the environment.

Like German socialist philosopher Karl Marx, Murray Bookchin argued that the human race cannot survive in a civilization based on life in the modern city, bureaucratic decision-making structures, and industrialized labor. Bookchin, however, built on Marx's insights to argue further that both socialism and capitalism are heedless of modern industry's impacts on the environment, and so neither socialism nor capitalism can be the basis for a sustainable society. He posited a close interconnection between human domination over nature and human beings' domination over one another, arguing that since the propensities to dominate nature and to dominate other humans sprang up together, they must be eliminated together.

Bookchin published his first two books, *Our Synthetic Environment* (1962) and *Crisis in Our Cities* (1965), under the pseudonym Lewis Herber. In these works he argued that it would not be an uprising of the proletariat but rather an uprising of an antiauthoritarian younger generation that would resolve the environmental crisis by dissolving all social hierarchies. This has not come to pass, of course, and some have criticized Bookchin's faith in the younger generation as naïve. His insight concerning the connection between social problems and the earth's environment, however—that the existence of hierarchy, in addition to the misuse of technology, has brought humankind to the brink of disaster—is seen as a valuable contribution to environmental thought.

Throughout his career, Bookchin developed and refined his argument that a link exists between the "destructive logic behind a hierarchical social structure" and environmental crisis. He strongly criticized environmentalists who are satisfied to save endangered species or ban harmful chemicals yet support underlying social structures that produce new toxins and similar problems. He called for a return to life as it was before the Industrial Revolution, in particular citing what he saw as the organic and harmonious way of life of past societies such as the Plains Indians in North America. Some critics also deemed this notion naïve.

In addition to his writing, Bookchin served as professor of social ecology at Ramapo College in New Jersey and as director of the Institute for Social Ecology at Goddard College in Rochester, Vermont. Among his books are *Toward an Ecological Society* (1980), *The Ecology of Freedom: The Emergence and Dissolution of Hierarchy* (1982), *Remaking Society* (1989), *The Philosophy of Social Ecology* (1990), *The Rise of Urbanization and the Decline of Citizenship* (1987), and *Social Ecology and Communalism* (2007).

Anne Statham

FURTHER READING

Barry, John. "Murray Bookchin, 1921-   ." In *Fifty Key Thinkers on the Environment*, edited by Joy A. Palmer. New York: Routledge, 2001.

Bookchin, Murray. "What Is Social Ecology?" In *Earth Ethics: Introductory Readings on Animal Rights and Environmental Ethics*, edited by James P. Sterba. 2d ed. Upper Saddle River, N.J.: Prentice Hall, 2000.

White, Damian. *Bookchin: A Critical Appraisal.* London: Pluto Press, 2008.

Borlaug, Norman

CATEGORIES: Activism and advocacy; agriculture and food

IDENTIFICATION: American plant pathologist and environmental activist

BORN: March 25, 1914; Cresco, Iowa

DIED: September 12, 2009; Dallas, Texas

SIGNIFICANCE: Borlaug, who became known as the father of the Green Revolution, pioneered efforts to develop high-yield crops to increase food production throughout the world.

Norman Borlaug credited his childhood experiences on his family's farm with providing him with a practical approach to agriculture. In the 1930's Borlaug studied forest management and plant pathology at the University of Minnesota, earning a doctorate by 1942. In 1943 the Rockefeller Foundation, which U.S. secretary of agriculture Henry Wallace had convinced to fund agricultural aid for Mexican farmers, hired Borlaug to breed disease-immune crops that could be grown in varied climates. He perfected a strain of high-yield dwarf spring wheat.

While Borlaug was experimenting with plant breeding, the post-World War II global population rapidly increased. Some environmentalists, such as Paul R. Ehrlich, predicted that not enough food could be produced and mass starvation would occur. Borlaug believed that his wheat could prevent such a disaster. In 1963 the Rockefeller Foundation and the Mexican government established the International Maize and Wheat Improvement Center (known as CIMMYT, for its name in Spanish, Centro Internacional de Mejoramiento de Maíz y Trigo) and named Borlaug director of the Wheat Improvement Program. He traveled to India and other developing nations to share his high-yield agricultural techniques. Critics thought that Borlaug should plant indigenous crops rather than Western grains, but he emphasized that wheat provides necessary calories and nutrients.

The Indian and Pakistani governments resisted Borlaug's efforts until famine in their countries became extreme. Delayed seed shipments and the outbreak of war between India and Pakistan also hindered Borlaug's work. He persisted, however, and yields increased approximately 70 percent during the first season. By 1968 Pakistan was agriculturally self-sufficient and increased its yields from 3.4 million tons of wheat in 1963 to 18 million by 1997. India boosted its yields from 11 million tons to 60 million tons, even briefly exporting wheat.

The expansion of food production spearheaded by Borlaug, which saved hundreds of millions of people from starvation, was called the Green Revolution. Borlaug was awarded the Nobel Peace Prize in 1970 for his humanitarian efforts to secure the basic human right of freedom from hunger. This honor did not ensure continued support of his work, however. When he expressed interest in using his techniques to assist African agriculture, Borlaug's CIMMYT patrons stopped his funding because of a backlash among environmentalists that protested high-yield agriculture, claiming that its use of inorganic fertilizers and irrigation damages the environment. Borlaug countered with the argument that high-yield agriculture preserves habitats from slash-and-burn techniques used to create farmland. He criticized theorists who, in his opinion, do not comprehend the reality of what is economically and politically possible in developing countries. With colleagues Haldore Hanson and R. Glenn Anderson, he wrote *Wheat in the Third World* (1982).

Borlaug's projects in African nations were funded by Sasakawa-Global 2000, which was financed by former U.S. president Jimmy Carter and Japanese industrialist Ryoichi Sasakawa. Continuing the development of high-yield crop strains at CIMMYT in the 1990's, Borlaug wrote articles, lectured, and testified before the U.S. Congress about the opposition to his work posed by environmental lobbyists. After receiving government support in Ethiopia and Poland, he planned similar agricultural programs in the former Soviet Union and Latin America. Warning that arable land is finite, Borlaug stressed that uncontrolled population growth could result in starvation in the twenty-first century and demanded the storage of food reserves.

Elizabeth D. Schafer

FURTHER READING

Borlaug, Norman. "Are We Going Mad?" In *The Ethics of Food: A Reader for the Twenty-first Century*, edited by Gregory E. Pence. Lanham, Md.: Rowman & Littlefield, 2002.

Hesser, Leon. *The Man Who Fed the World: Nobel Peace Prize Laureate Norman Borlaug and His Battle to End World Hunger.* Dallas: Durban House, 2006.

Paarlberg, Robert. *Starved for Science: How Biotechnology Is Being Kept out of Africa.* Cambridge, Mass.: Harvard University Press, 2008.

Brockovich, Erin

CATEGORIES: Activism and advocacy; human health and the environment
IDENTIFICATION: Legal clerk and environmental activist
BORN: June 22, 1960; Lawrence, Kansas
SIGNIFICANCE: Brockovich helped construct a legal case against the Pacific Gas and Electric Company for its role in polluting the drinking water of Hinkley, California, with chromium 6. The clients in the case received the largest settlement ever made in the United States in a direct-action lawsuit.

Erin Brockovich was born Erin Pattee; her mother, Betty Jo O'Neal-Pattee, was a journalist, and her father, Frank Pattee, was an industrial engineer. She graduated from Lawrence High School in 1978 and earned an associate in applied arts degree from Wade Business College in Dallas in 1980. She then became a management trainee for the Kmart retail chain and moved to California. After a period in which she married and divorced twice (Steve Brockovich was her second husband), had three children, and held a variety of jobs, she became a secretary for the law firm of Masry & Vititoe of Northridge, California, in 1991.

Brockovich was assigned to work on a pro bono case for a home owner in Hinkley, California, and quickly developed a rapport with the client, who was bringing a lawsuit that alleged contamination of the town's drinking water with chromium 6. Residents of the area, it was alleged, were experiencing above-average numbers of miscarriages and cancers. Brockovich discovered that from 1952 to 1966 the Pacific Gas and Electric Company (PG&E) used chromium 6 to fight corrosion in the cooling tower of its natural gas pumping station in Hinkley. Wastewater containing chromium 6 was then pumped into unlined ponds, from which it leached into the aquifer that supplied Hinkley's drinking water. Brockovich pursued the case tenaciously and signed up more than six hundred additional persons to participate in the lawsuit. In 1996, after winning an initial round in court, the clients agreed to PG&E's offer to settle all their claims against the company for $333 million—a record high settlement for a direct-action lawsuit in the United States.

Thomas R. Feller

Brower, David

CATEGORIES: Activism and advocacy; preservation and wilderness issues
IDENTIFICATION: American environmental activist and writer
BORN: July 1, 1912; Berkeley, California
DIED: November 5, 2000; Berkeley, California
SIGNIFICANCE: Brower, who was vigorously involved in battles concerning environmental issues for more than fifty years, was one of the twentieth century's most influential and controversial environmental activists and writers.

David Ross Brower served as the first executive director of the Sierra Club from 1952 to 1969. He is credited by many with helping the San Francisco-based organization grow from two thousand to seventy-seven thousand members and developing it into a powerful national organization. He led the club in aggressive campaigns against U.S. government projects to develop wild areas, most notably fights that successfully stopped the construction of the Echo Park Dam, which would have flooded part of Dinosaur National Monument in Utah in the 1950's, and two different dams across the Colorado River in the Grand Canyon in the 1960's and 1970's. His enterprising tactics included full-page advertisements in *The New York Times* and the *San Francisco Chronicle*, which resulted in the Internal Revenue Service reclassifying the nonprofit Sierra Club as a lobbying organization and removing its tax-deductible status.

For more than twenty-five years, Brower focused much of his passion and energy on the Glen Canyon Dam in northeastern Arizona. "Glen Canyon died, and I was partly responsible for its needless death," Brower wrote in Eliot Porter's *The Place No One Knew: Glen Canyon on the Colorado* (1963). In the mid-1950's the U.S. Bureau of Reclamation was planning to construct dams across the Colorado River in the Grand Canyon and Glen Canyon, Arizona, and across the Green and Yampa rivers in Utah. Following the directives of the Sierra Club board of directors, Brower agreed to drop the club's opposition to the Colorado River dams if the Bureau of Reclamation would discontinue plans to build the two dams in Utah. The bureau agreed to the deal and moved forward to build the Glen Canyon Dam.

Before the dam construction was completed, Brower and the Sierra Club decided the compromise had been a mistake. They blamed their decision on a lack of familiarity with the spectacular beauty of Glen Canyon. (In 1996 the Sierra Club directors unanimously passed a motion by Brower to support draining Lake Powell, the reservoir behind the Glen Canyon Dam, and return the Colorado River flow to the most natural state possible. Brower did not advocate dismantling the dam; rather, he stated, it should be left "as a tourist attraction, like the Pyramids, with passers-by wondering how humanity ever built it, and why.")

In the mid-1960's Brower and the Sierra Club successfully led an effort to prevent construction of the Bureau of Reclamation's proposed Marble Canyon Dam in the Grand Canyon and helped cripple the bureau's effort to build Bridge Canyon Dam farther downstream in the Grand Canyon. By 1969, however, the majority of the Sierra Club's board of directors found Brower's tactics too reckless, both financially and politically, and they removed him as executive director. He then formed the preservation-oriented Friends of the Earth and the League of Conservation Voters, both of which flourished under his leadership. He also facilitated the establishment of independent Friends of the Earth organizations in other countries.

In 1982, after conflicts with members of the Friends of the Earth's professional staff and its directors, Brower moved on to form another group, Earth Island Institute, the stated mission of which was to globalize the environmental movement. He returned to the Sierra Club as a director in 1983 and was reelected in 1986 and 1995. In the fall of 1994 Brower helped develop the Ecological Council of Americas to improve cooperation among organizations in the Western Hemisphere that were attempting to integrate environmental and economic needs.

In the 1990's Brower called on the federal government to replace the U.S. Bureau of Land Management with a new agency called the National Land Service. Its mission would be to protect and restore private and public land in the United States. He also strongly advocated the creation of a national biosphere reserve system.

Throughout his life, Brower pushed the edges of environmental thought of the day. He pioneered ideas and methods to preserve the environment and create a global approach to issues. For many years Brower advocated the establishment of international natural reserves in areas of rich biodiversity and ecosystems. The United Nations Educational, Scientific, and Cultural Organization (UNESCO) has established such a system of World Heritage Sites.

Brower also advocated a method known as CPR to guard against the destruction of natural areas and biodiversity: "C" is for *conservation*, or the rational use of resources: "P" represents the *preservation* of threatened, endangered, and yet undiscovered species; and "R" stands for *restoration* of lands already damaged by human activities. Many of his tactics and ideas seemed radical when he introduced them but later became standard practice among mainstream environmentalists. Russell Train, chairman of the Council on Environmental Quality during President

Richard Nixon's administration, once said, "Thank God for Dave Brower; he makes it so easy for the rest of us to be reasonable."

Louise D. Hose

FURTHER READING

Brower, David. *Let the Mountains Talk, Let the Rivers Run: A Call to Save the Earth.* New ed. San Francisco: Sierra Club Books, 2007.

_____. "The Sermon." In *Speaking of Earth: Environmental Speeches That Moved the World*, edited by Alon Tal. New Brunswick, N.J.: Rutgers University Press, 2006.

McPhee, John. *Encounters with the Archdruid.* 1971. Reprint. New York: Farrar, Straus and Giroux, 2000.

Porter, Eliot. *The Place No One Knew: Glen Canyon on the Colorado.* Edited by David Brower. Commemorative ed. Layton, Utah: Gibbs Smith, 2000.

Stoll, Steven. *U.S. Environmentalism Since 1945: A Brief History with Documents.* New York: Palgrave Macmillan, 2007.

Brown, Lester

CATEGORIES: Activism and advocacy; agriculture and food
IDENTIFICATION: American agricultural scientist and author
BORN: March 28, 1934; Bridgeton, New Jersey
SIGNIFICANCE: Brown founded the Worldwatch Institute, an environmental think tank the mission of which is to analyze the state of the earth and to act as "a global early warning system."

Lester Brown was raised on a small tomato farm in Bridgeton, New Jersey. He joined the 4-H Club and the Future Farmers of America at his local school. When he was fourteen, he and his brother purchased a used tractor and a small plot of land to grow tomatoes. Within a brief time, they became two of the most successful tomato farmers on the East Coast. Brown graduated from Rutgers University in 1955 with a degree in agricultural science and immediately put his education to practical use. He worked for six months in a small farming community in India, becoming intimately acquainted with hunger problems created by population growth and unsustainable agricultural practices.

In 1959 Brown earned a master's degree in agricultural economics and soon after joined the U.S. Department of Agriculture as an international agricultural analyst. After leaving that post in 1969, he helped organize the Overseas Development Council, a private group devoted to analyzing issues relevant to relations between the United States and developing countries.

Brown's educational and career background prepared him well for his work and leadership in the Worldwatch Institute. While living and working in developing nations, he became acutely aware of such problems as the extensive poverty caused by economic systems dependent on cash crops for export to wealthy industrial countries and the use of agricultural practices that cause deforestation and desertification. He realized that food security could replace military security as the major concern of governments in the twenty-first century.

Shortly after Brown established the Worldwatch Institute in 1974, he and other staff members initiated the Worldwatch Papers, which focus on population growth and the resulting stress on natural resources, transportation trends, and the human and environmental impact of urbanization. In 1984 Brown established the institute's annual report, *State of the World*, a comprehensive overview of specific global environmental issues. This publication, now available in more than twenty-five languages, is used by political leaders, educators, and citizens as a resource for information on environmental problems and ways to address them. In 1992 Brown inaugurated the publication of *Vital Signs: The Trends That Are Shaping Our Future.* This annual handbook features environmental, economic, and social statistical indicators on trends likely to influence the world's future.

In Brown's view, global environmental problems should be addressed through international efforts funded by taxes on currency exchanges, taxes for pollution emissions, and a greater involvement by the United Nations. He has advocated a shift away from national spending on unsustainable economic growth and toward investing in research and development that enhances environmental quality and protection of natural resources.

Brown has published numerous books, some in partnership with the Worldwatch Institute and some published by the Earth Policy Institute, which Brown founded in 2001 to raise awareness of environmental issues. His books include *In the Human Interest: A Strategy to Stabilize World Population* (1974), *Building a Sus-*

tainable Society (1981), *Who Will Feed China? Wake-Up Call for a Small Planet* (1995), and *Plan B: Rescuing a Planet Under Stress and a Civilization in Trouble* (2003). In these and other writings, Brown warns about the ecological dangers of overexploiting the earth's resources and urges those in the world's developed nations to change their lifestyles. He also advises governments and scientists to cooperate in finding solutions to environmental problems.

Ruth Bamberger

FURTHER READING

Brown, Lester. "Worldwatch." In *Life Stories: World-Renowned Scientists Reflect on Their Lives and on the Future of Life on Earth*, edited by Heather Newbold. Berkeley: University of California Press, 2000.

Nelson, David E. "In Praise of Lester Brown." *Futurist* 42, no. 6 (2008).

Wallis, Victor. "Lester Brown, the Worldwatch Institute, and the Dilemmas of Technocratic Revolution." *Organization and Environment* 10, no. 2 (1997): 109-125.

Bureau of Land Management, U.S.

CATEGORIES: Organizations and agencies; land and land use

IDENTIFICATION: Federal agency responsible for managing the public lands of the United States

DATE: Established in 1946

SIGNIFICANCE: The Bureau of Land Management is charged with managing public lands in ways that are consistent with both multiple-use concepts and sustained-yield principles.

The U.S. federal government's original policy concerning public lands was to encourage their disposal. The most widely known method for this disposal was through homesteading as a result of the Homestead Act of 1862, which was overseen by the General Land Office, a forerunner of the Bureau of Land Management (BLM) created in 1812. The land that remained after settlement and the designation of national parks, national forests, and wildlife refuges was available for public use. Abuses of these public lands became widespread, however, and by the 1930's there was need for correction. Extensive overgrazing of livestock constituted one of the most serious misuses of public lands. As a result of these abuses, the Grazing Service was created as part of the U.S. Department of the Interior to manage some 32.4 million hectares (80 million acres) under the provisions of the Taylor Grazing Act in 1934.

In 1946 the Grazing Service became the BLM. Part of the BLM's continuing responsibilities were based on the need to evaluate damage, classify public lands for grazing purposes, and assess fees for grazing. Concerns about environmental quality grew during the 1960's and 1970's, and these increased concerns extended to the public lands. As a result, the BLM was granted more authority under the provisions of the Federal Land Policy and Management Act of 1976, which encouraged the BLM to manage public lands in ways that are consistent with both multiple-use concepts and sustained-yield principles.

The multiple-use approach to land-use planning has a lengthy history in resource management in the United States, particularly in the forestry area. Because of the potential for land to be used for a wide variety of purposes, such as timber production, grazing, and recreation, legislators recognized that careful planning is needed and that there are strong advantages to managing public lands in a way that ensures that resources are sustained and the environment is protected. These principles were articulated in the Multiple Use-Sustained Yield Act of 1960, and they have, in turn, become a part of BLM policy.

In the twenty-first century, most lands managed by the BLM are in Alaska and the other states west of the Mississippi River. However, the management of onshore oil drilling, gas production, and mineral development on federal lands is also part of the BLM's responsibilities. As a result, the bureau maintains an office to deal with oil and mineral policies on public land east of the Mississippi River.

Jerry E. Green

FURTHER READING

Allen, Leslie. *Wildlands of the West: The Story of the Bureau of Land Management.* Washington, D.C.: National Geographic Society, 2002.

Skillen, James. *The Nation's Largest Landlord: The Bureau of Land Management in the American West.* Lawrence: University Press of Kansas, 2009.

Burroughs, John

CATEGORIES: Activism and advocacy; preservation and wilderness issues
IDENTIFICATION: American nature writer
BORN: April 3, 1837; near Roxbury, New York
DIED: March 29, 1921; en route from California to New York
SIGNIFICANCE: Through his best-selling books, Burroughs raised Americans' awareness of the beauty of nature and the importance of preserving it.

John Burroughs grew up on a farm in the Catskill Mountains in New York, spending as much time as he could outdoors. Around the age of twenty, he decided that he would try to earn his living as a writer. After a brief teaching career, he spent ten years as a clerk in the U.S. Treasury Department in Washington, D.C. As a sideline, he published magazine essays about natural history and philosophy, always working to sharpen his writing skills. During these years he published his first book, *Notes on Walt Whitman as Poet and Person* (1867), the first biography of the great poet, who had also been a government clerk and who was Burroughs's personal friend.

Burroughs's first volume of essays, *Wake-Robin* (1871), was representative of the twenty-two collections that would follow: It featured close observations of natural history and commentary about simple country life, was made up mostly of essays (including such titles as "Birds' Nests" and "In the Hemlocks") that had been previously published in magazines, and won immediate acclaim. To Burroughs's first readers, the genre of nature writing was new and captivating, and Burroughs soon became its most popular practitioner. Sure now that he could live by his pen, he left his government job and moved back to New York, establishing a small fruit farm on the banks of the Hudson River in 1873. He continued to publish essays in

John Burroughs, right, poses in Alaska in 1899 with naturalist and preservationist John Muir, one of the many admirers of Burroughs's nature writings. (Library of Congress)

some of the most popular magazines of his day and collected them into new books approximately every two years. His titles reveal something of the simple wonder that informs these books: *Fresh Fields* (1885), *Bird and Bough* (1906), and *Under the Apple-Trees* (1916).

Burroughs's work remained popular throughout his lifetime—a rare achievement for a writer. He formed lasting friendships with many of his admirers, including John Muir, Thomas Edison, and Henry Ford. Two more friendships led to books: *John James Audubon* (1902), an appreciation and biography, and *Camping with President Roosevelt* (1906).

After Burroughs's death in 1921, the John Burroughs Association was founded and the John Burroughs Sanctuary was established to preserve his property and many of his books in West Park, New York. Burroughs continues to be acknowledged as a pioneer and a master of the genre of nature writing, and many of his books remain in print. The John Burroughs Association presents two annual awards for nature writing published in the preceding year: one to the author of an outstanding natural history essay and one to the author of a distinguished book of natural history. In 1997 Burroughs was named a charter member of the Ecology Hall of Fame in Santa Cruz, California.

Cynthia A. Bily

FURTHER READING

Walker, Charlotte Zoë, ed. *Sharp Eyes: John Burroughs and American Nature Writing.* Syracuse, N.Y.: Syracuse University Press, 2000.

Warren, James Perrin. *John Burroughs and the Place of Nature.* Athens: University of Georgia Press, 2006.

Ceres

CATEGORIES: Organizations and agencies; activism and advocacy
IDENTIFICATION: American nonprofit organization dedicated to injecting environmental considerations into investment decisions
DATE: Established in 1989
SIGNIFICANCE: Ceres promotes the well-being of the environment and human society by uniting investors and key stakeholders, environmental organizations, and public interest groups to work with corporations and capital markets to integrate sustainability and environmentally responsible practices into their day-to-day operations.

The 1989 *Exxon Valdez* oil spill in Prince William Sound was the catalyst for the formation of Ceres. Meeting in Chapel Hill, North Carolina, fifteen major environmental groups joined with investors and public pension fund managers to encourage greater corporate consideration of the environmental consequences of their actions. With participation by the New York and California state pension funds, the Social Investment Forum, and a coalition of more than two hundred Protestant and Roman Catholic groups, the initial members of that group represented more than $150 billion in invested capital. The group was originally known by the acronym CERES (for Ceres, the Roman goddess of agriculture), which stood for the Coalition for Environmentally Responsible Economies; it eventually dropped the longer name altogether and became simply Ceres.

Ceres was founded on the notion that government regulation alone is not sufficient to effect material improvements in the environment. Rather, progress requires fundamental changes in corporate behavior and more responsible attitudes toward the environment. According to Ceres, investors can influence this shift in beneficial ways if their investment decisions include consideration of corporations' environmental track records. Ceres's first project was the development of a corporate code of environmental ethics. Originally called the Valdez Principles as a reminder of the oil spill, the code later become known as the Ceres Principles. This ten-point code of conduct calls for corporations to protect the biosphere, use natural resources sustainably, reduce and responsibly dispose of wastes, conserve energy, reduce environmental and health risks to employees and communities, offer safe products and services, restore the environment, inform the public of hazards, ensure management commitment to environmental responsibility, and issue audit reports.

Ceres uses shareholder resolutions to initiate discussions of environmental responsibility at the highest corporate levels. In some corporations, such resolutions eventually lead to formal endorsement of the ten Ceres Principles. By endorsing the principles, companies acknowledge their environmental responsibility, actively commit to an ongoing process of continuous improvement, and agree to initiate comprehensive public reporting of environmental issues.

Initially, the Ceres Principles were adopted by companies that already had strong "green" reputations. In 1993, following lengthy negotiations, Sunoco became the first *Fortune* 500 company to endorse the principles. Several other large companies, including Bank-Boston, Bethlehem Steel, Coca-Cola, General Motors, and Polaroid, soon followed Sunoco's example. By 2010 more than fifty companies had endorsed the Ceres Principles, including thirteen *Fortune* 500 companies that had adopted their own equivalent environmental codes of conduct.

In 1997 Ceres launched the Global Reporting Initiative (GRI), which standardizes corporate environmental reporting to generate the equivalent of a financial report. Just as an official corporate financial report includes required information on expenses, revenue, and profitability, each company's GRI report includes required information on sustainability, environmental and social impacts, technological innovation, and other pertinent issues. GRI reports improve corporate accountability by ensuring that all stakeholders—investors, fund managers, community groups, environmentalists, and labor organizations—have access to standardized and consistent information. Armed with these data, environmentally conscious investors can measure corporate compliance with the Ceres Principles and thus use capital markets effectively to promote sustainable business practices.

Ceres partnered with the United Nations Environment Programme to issue draft GRI sustainability reporting guidelines in 1999. Finalized guidelines were released in 2000, and fifty organizations responded with their reports. In 2002 the GRI was established as an independent international body based in the Netherlands, and in 2009 more than twelve hundred organizations issued GRI reports.

Allan Jenkins
Updated by Karen N. Kähler

FURTHER READING

Ceres. *Twenty-first Century Corporation: The Ceres Roadmap for Sustainability.* Boston: Author, 2010.

Friedman, Frank B. *Practical Guide to Environmental Management.* 10th ed. Washington, D.C.: Environmental Law Institute, 2006.

Leipziger, Deborah. *The Corporate Responsibility Code Book.* Sheffield, England: Greenleaf, 2003.

Chipko Andolan movement

CATEGORIES: Organizations and agencies; activism and advocacy; forests and plants

IDENTIFICATION: Movement started by villagers in northern India to stop lumber companies from clear-cutting mountain slopes

DATE: Originated in April, 1973

SIGNIFICANCE: Through nonviolent protest, the Chipko Andolan movement put pressure on the Indian government to develop policies concerning natural resources that would be sensitive to the environment and to the needs of all the Indian people.

The forests of India are a critical resource for the subsistence of rural people throughout the country, especially in the hill and mountain areas. Mountain villagers depend on the forests for firewood, for fodder for their cattle, for wood for their houses and farm tools, and as a means to stabilize their water and soil resources. During the 1960's and 1970's the Indian government restricted villagers from huge areas of forestland, then auctioned off the trees to lumber companies and industries located in the plains. Large

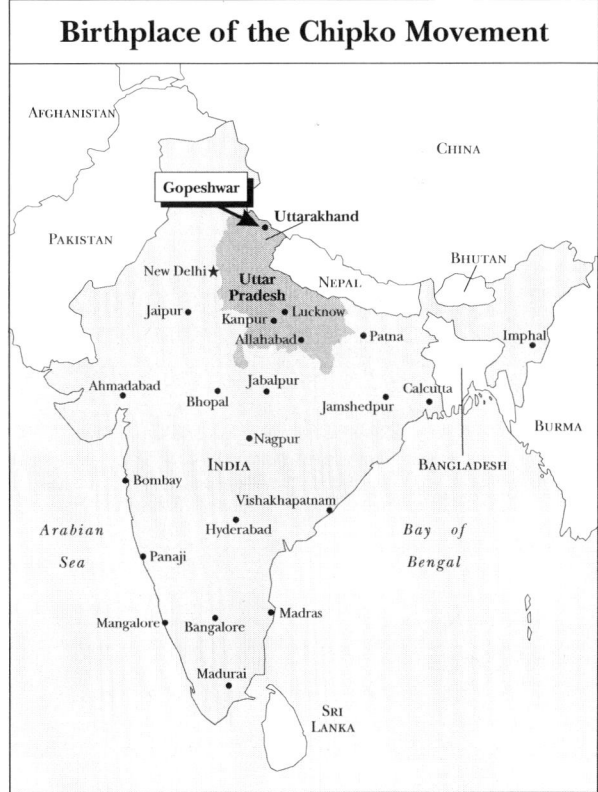

Birthplace of the Chipko Movement

lots of trees were sold to the highest bidders by the Forest Department, with the purchase payments going to the Indian government.

Because of government restrictions and an ever-growing population, women who lived in mountain villages were forced to walk for hours each day just to gather firewood and fodder. In addition, when mountain slopes were cleared of trees, rains washed away the topsoil, leaving the soil and rocks underneath to crumble and fall in landslides. Much of the soil from the mountain slopes was deposited in the rivers below, raising water levels. At the same time, the bare slopes allowed much more rain to run off directly into the rivers, which resulted in flooding.

As trees were being felled for commerce and industry at increasing rates during the early 1970's, Indian villagers finally sought to protect their lands and livelihoods through a method of nonviolent resistance inspired by Indian leader Mohandas Gandhi. In 1973 this resistance spread and became organized into the Chipko Andolan movement, commonly referred to as the Chipko movement. The word *chipko* comes from a Hindi word meaning "to embrace," while *andolan* refers to a protest against harmful practices. Together the words literally mean "movement to hug trees."

The movement originated in April, 1973, as a spontaneous protest by mountain villagers against logging abuses in Uttar Pradesh, an Indian province in the Himalayas. When contractors sent their workers in to fell the trees, the villagers embraced the trees, saving them by interposing their bodies between the trees and the workers' axes. The movement was largely organized and orchestrated by village women, who became leaders and activists in order to save their means of subsistence and their communities.

After many Chipko protests in Uttar Pradesh, victory was finally achieved in 1980 when the Indian government placed a fifteen-year ban on felling live trees in the Himalayan forests. The Chipko movement soon spread to other parts of India, and clear-cutting was stopped in the Western Ghats and the Vindhya Range.

The Chipko protesters staged a socioeconomic revolution in India by gaining control of forest resources from the hands of a distant government bureaucracy that was concerned only with selling the forest in order to make urban-oriented products. The movement generated pressure for the Indian government to develop a natural resources policy that was more sensitive to the environment and the needs of all people.

Alvin K. Benson

FURTHER READING

Guha, Ramachandra. "Chipko: Social History of an 'Environmental' Movement." In *Social Movements and the State*, edited by Ghanshyam Shah. Thousand Oaks, Calif.: Sage, 2003.

Hill, Christopher V. *South Asia: An Environmental History.* Santa Barbara, Calif.: ABC-CLIO, 2008.

Cloning

CATEGORY: Biotechnology and genetic engineering
DEFINITION: Production of a population of genetically identical cells or organisms derived from a single ancestor, or the production and amplification of identical deoxyribonucleic acid molecules
SIGNIFICANCE: Gene cloning is commonly used to produce genetically modified organisms and genetically engineered crops. Reproductive cloning has the potential to be used to mass-reproduce animals with special qualities, or to expand the population of an endangered species. With further development, therapeutic cloning might someday be employed to produce whole organs from single cells or replace disease-damaged cells with healthy ones.

The molecular cloning and engineering of deoxyribonucleic acid (DNA) molecules were first made possible with the discoveries of DNA ligase (enzymes that join DNA molecules) in 1967 and restriction endonucleases (enzymes that cut DNA molecules at specific nucleotide sequences) in 1970 by Hamilton Smith and Daniel Nathans. These enzymes allow scientists to cut and join DNA molecules from different species to produce recombinant DNA. For example, the DNA encoding human insulin can be combined with a plasmid, a small piece of DNA often found in bacteria such as *Escherichia coli* (*E. coli*). After the recombinant human-insulin-DNA/plasmid-DNA molecule is constructed, it can be inserted into a host cell such as *E. coli*. The recombinant DNA molecule will then replicate one or more times each time the *E. coli* DNA replicates. Thus a clone of identical recombinant human-insulin-DNA/plasmid-DNA molecules will result. If the recombinant molecule has been engineered with the requisite signals, the *E. coli* will produce copious amounts of human insulin. In 1972 the first recombinant DNA molecules were made at Stanford University, and in 1973 such molecules were inserted via plasmids into *E. coli*. The first successful synthesis of a human protein by *E. coli* was somatostatin, reported in 1977 by Keiichi Itakura and coworkers. In 1984 insulin was the first human protein made by *E. coli* to become commercially available.

The production of cloned recombinant DNA has been an indispensable tool in biological research and has increased the knowledge and understanding of the structure and function of DNA and the control of gene activity. Cloned DNA has led to the manufacture of important products of medical interest, the production of DNA for gene transfer and genetic engineering experiments, and the identification of mutations and genetic disease. Several different vectors have been developed for the delivery of DNA to a variety of plant, animal, and protistan cells, resulting in the creation of many transgenic species. Transgenic plants have been produced that are resistant to certain herbicides, insects, and viruses, and a variety of animals have been engineered with the human growth factor gene. Transgenic plants, animals, and protistans now produce a variety of human proteins, including insulin, antithrombin, growth hormone, clotting factor, vaccines, and many other pharmaceuticals and therapeutic agents, as well as molecular probes for the diagnosis of human disease. As a result of this technology, the costs of treating many diseases have declined.

The first human gene transfer experiment using cloned DNA was performed in 1990 on a four-year-old girl with severe combined immune deficiency (SCID). SCID is caused by a mutation in the adenine deaminase (ADA) gene that results in white blood cells deficient in their immune response. These cells were removed from the girl, the normal gene was inserted, and the genetically altered white cells were returned to her body, where they repopulated and expressed normal defense mechanisms. The girl's white cell numbers normalized after repeated treatments. While she continued to require ADA enzyme injections for primary management of her condition, she did develop normal immunity over time.

ENVIRONMENTAL IMPACT OF DNA CLONING

The construction of recombinant DNA molecules and their subsequent cloning have not been without controversy. In 1971 researcher Paul Berg planned an experiment to combine DNA from simian virus 40 (SV40)—a virus that causes tumors in monkeys and transforms human cells in culture—with bacteriophage *l* and to incorporate the recombinant molecule into *E. coli*. However, several scientists warned of a potential biohazard. Because *E. coli* is a natural inhabitant of the human digestive system, it was feared that the engineered *E. coli* could escape from the laboratory, enter the environment, become ingested by humans, and cause cancer as a result of its newly acquired DNA. The scientific community imposed a moratorium on recombinant DNA work in 1974 until the National Institutes of Health (NIH) could study the safety of recombinant DNA research and develop guidelines under which such work could proceed. The guidelines, originally published in 1976, were eventually relaxed after it was clearly demonstrated that the work was not nearly as dangerous as initially feared.

In 1983 the NIH granted permission to the University of California at Berkeley to release bacteria that had been engineered to protect plants from frost damage. This was the first experiment intentionally designed to introduce genetically engineered organisms into the environment. Various environmental and consumer protection groups were successful in persuading federal judge John J. Sirica to order the suspension of the planned trial. Environmentalists feared that the bacteria could spread to other plants, prolong the growing season, cause irreparable harm to the environment, or enter the atmosphere and decrease cloud formation or alter the climate. The NIH and the university eventually won approval for their experiment, and the field trials were carried out in 1987.

While many scientists believe that the introduction of tested and approved engineered species into the environment will prove harmless and that early concerns were unfounded, many environmentalists still have reservations because the long-term environmental effects of genetically altered organisms remain unknown. It is feared that the introduction of some organisms will have negative impacts on the environment and irreversible global effects. Many worry that a genetically engineered organism such as a bacterium or virus could spread throughout the environment, causing human disease or ecological destruction.

The use of genetically engineered organisms could have many adverse effects. The introduction of new genes into an organism could extend the range of that species, causing it to infringe on the natural habitats

Milestones in Cloning

Year	Event
1892	Hans Adolph Eduard Dreisch clones sea urchins by separating the first two and four blastomeres.
1902	Hans Spemann successfully repeats Dreisch's experiment using salamanders.
1952	Robert Briggs and Thomas J. King successfully clone frogs by nuclear transplantation of embryonic nuclei to enucleated eggs.
1967	Deoxyribonucleic acid (DNA) ligase, the enzyme that joins DNA molecules, is discovered.
1969-1970	Daniel Nathans, Hamilton Smith, and others discover restriction endonucleases.
1971	Paul Berg plans to combine DNA with a bacteriophage and insert the recombinant DNA molecule into *Escherichia coli* (*E. coli*).
1972	The first recombinant DNA molecules are constructed at Stanford University.
1973	Recombinant DNA is first inserted into *E. coli*.
1974	Scientists call for a moratorium on recombinant DNA research until the National Institutes of Health (NIH) can study the safety of recombinant DNA research and develop guidelines.
1976	The NIH issues its guidelines for recombinant DNA research.
1983	The NIH grants permission to the University of California at Berkeley to release genetically engineered bacteria designed to retard frost formation on plants.
1986	Steen Willadsen clones sheep using early embryonic nuclei.
1987	Randall Prather and Willard Eyestone clone cows using early embryonic nuclei.
1990	The first human gene transfer experiment is performed on a patient with severe combined immune deficiency (SCID).
1997	Ian Wilmut and Keith Campbell announce the successful cloning of Dolly the sheep, the first mammal cloned from an adult cell.
2001	The first endangered wild animal, an Asian ox (gaur) is cloned.
2002	A study finds that the genes in about 4 percent of cloned mice function abnormally.
2003	Dolly the sheep is euthanized after contracting lung disease.
2009	The first clone of an extinct animal, a Pyrenean ibex (a species of wild goat), is born but dies within minutes.

of closely related or more distant species and thus disrupt the balance of nature. Many examples already exist of ecological disruption that has occurred as a result of the introduction of plant and animal species into areas where they have no natural predators. Some environmentalists believe that the use of herbicide-resistant crops will serve to prolong, extend, and even increase the use of toxic herbicides. Others fear that herbicide-resistant genes could be transferred to related plants, producing a population of herbicide-resistant weeds.

On the other hand, the introduction of organisms engineered through the cloning of recombinant DNA into the environment could have significant positive environmental impacts and increase world food production. Genetically altered plants could reduce the input of toxic chemicals into the environment. Genetic modification can create crops with larger yields; resistance to pests, pesticides, drought, and disease; higher tolerance to cold, heat, and drought; longer shelf life; and greater photosynthetic and nitrogen-fixing activity. Plants genetically engineered to produce a natural insecticide that is nontoxic to humans ideally alleviate the need for insecticide application; with the toxin produced by and confined to the plants themselves, there is no need to contaminate the entire area where they are grown. Bioremediation of toxic waste dumps, chemical spills, and oil spills could be enhanced by genetically engineered microorganisms.

Animal and Plant Cloning

The cloning of plants from cuttings has been successfully practiced for thousands of years and is commonly used for many important food crops. Successful animal cloning was first reported in 1892 by Hans Adolph Eduard Dreisch. Dreisch separated the first two and four embryonic cells (blastomeres) of the sea urchin and allowed them to develop into complete, genetically identical embryos.

The first report of successful animal cloning by nuclear transplantation was published in 1952 by Robert Briggs and Thomas J. King, who removed nuclei from embryonic frog cells and transplanted them into eggs from which their nuclei had been removed. By pricking the eggs with a glass needle the scientists induced them to divide and often develop into complete tadpoles. The first reliable reports of successful animal cloning by nuclear transplantation in mammals came in 1986 from Steen Willadsen in Cambridge, England. Willadsen cloned sheep from the nucleus of an early blastula cell. In 1987 Randall Prather and Willard Eyestone cloned cows while working in Neal First's laboratory at the University of Wisconsin.

In 1997 Ian Wilmut and Keith Campbell announced that they had successfully cloned a sheep named Dolly in Edinburgh, Scotland, the year before. Dolly was a milestone in cloning research because she was the first mammal cloned from an adult cell. Such animal cloning makes genetic engineering more efficient, because an animal would have to be engineered only once and then could be used to donate nuclei for cloning. The use of adult cells in cloning is advantageous in that the donor animal's physical characteristics are known before the animal is cloned. However, clones created from adult cells share those cells' shortened telomeres (the sequences at the ends of chromosomes that grow shorter with each generation of cell replication). Telomere shortening is associated with the aging and death of cells and of the entire organism, which means that the cloned animal is likely to have a greater susceptibility to degenerative conditions and a shortened life expectancy. Dolly, who was cloned from a six-year-old sheep, lived only six years before she had to be euthanized in 2003 because of a progressive lung disease. (Her breed typically lives about twelve years.) She had developed arthritis the year before. Whether her illnesses stemmed from premature aging is unclear; however, her telomeres were found to be short.

Although inbred stocks have been used for centuries in agriculture, it is feared that the extensive use of animal and plant clones could severely reduce the genetic variability of various important crop, forestry, and livestock species, making them more susceptible to disease and extreme environmental factors. On the other hand, the development of successful cloning techniques could rescue endangered species from the brink of extinction or allow for the creation of a population of genetically identical animals, which would be valuable in medical research. Moreover, cloning could be used to create an entire population of animals from one individual that has been genetically altered to produce a valuable pharmaceutical or organs suitable for human transplant. However, controversy would undoubtedly arise regarding the ethics of mass-producing animals through cloning for this purpose.

Reproductive cloning is costly, and more than 90 percent of cloning attempts do not result in viable offspring. Cloned mammals have tended to have immune function problems, increased infection rates, susceptibility to tumors, and short lives. A 2002 study of cloned mice found that about 4 percent of their genes functioned abnormally. Members of the American Medical Association and the American Association for the Advancement of Science have issued formal public statements advising against reproductive cloning in humans.

Charles L. Vigue
Updated by Karen N. Kähler

Further Reading

Drlica, Karl. *Understanding DNA and Gene Cloning: A Guide for the Curious.* 4th ed. Hoboken, N.J.: John Wiley & Sons, 2004.

Fritz, Sandy, ed. *Understanding Cloning.* New York: Warner Books, 2002.

Klotzko, Arlene Judith. *A Clone of Your Own? The Science and Ethics of Cloning.* New York: Cambridge University Press, 2006.

Kolata, Gina Bari. *Clone: The Road to Dolly, and the Path Ahead.* New York: HarperCollins, 1998.

Williams, J. G., A. Ceccarelli, and A. Wallace. *Genetic Engineering.* 2d ed. Oxford, England: Bios, 2001.

Wilmut, Ian, Keith Campbell, and Colin Tudge. *The*

Second Creation: Dolly and the Age of Biological Control. Cambridge, Mass.: Harvard University Press, 2001.

Wilmut, Ian, and Roger Highfield. *After Dolly: The Uses and Misuses of Human Cloning.* New York: W. W. Norton, 2006.

Commoner, Barry

CATEGORIES: Activism and advocacy; nuclear power and radiation; energy and energy use
IDENTIFICATION: American biologist and antinuclear activist
BORN: May 28, 1917; Brooklyn, New York
SIGNIFICANCE: Commoner has raised public awareness of a number of important environmental issues, particularly regarding the use of energy resources, organic farming and pesticides, waste management, and toxic chemicals.

Barry Commoner graduated from Columbia University in 1937 and received an M.A. (1938) and a Ph.D. (1941) from Harvard University. He has been awarded a number of honorary degrees from prominent institutions, and his biography has appeared in *Who's Who in Science and Engineering.* His interest in becoming an environmental activist was sparked in the 1950's by the testing of nuclear weapons. In 1965 Commoner was president of the St. Louis Commission on Nuclear Information, and in 1966 he founded the Center for the Biology of Natural Systems (CBNS) at Washington University in St. Louis, Missouri. In 1981 the CBNS moved to Queens College, in the City University of New York (CUNY) system. The CBNS program has continually informed the public and government on environmental issues, including waste recycling and energy resources; it has provided extensive analysis of the generation and fate of dioxins.

Commoner is the author of several books on the environment, including *Science and Survival* (1966), *The Closing Circle: Nature, Man, and Technology* (1971), *Ecology and Social Action* (1973), *The Politics of Energy* (1979), and *Making Peace with the Planet* (1990). He publicizes environmental problems and relates them to modern technology. Commoner claims that modern methods of production cause human illness and that modern technologies that have resulted in environmental crises have safer alternatives. He believes that to restore the environment, humans need to adopt more benign ways of accomplishing tasks, such as eliminating the use of plastic wrap and recycling instead of incineration. He has promoted the view that there is economic value and profitability in the replacement technologies. He has drawn attention to the affiliations of some environmental scientists and has questioned their opinions, stating that environmental hazards are reliably recognized only through studies done by independent scientists. He has encouraged the public to become more informed, to achieve a greater understanding of the effects of particular technologies on the environment before using those technologies.

In recognition of his contributions to public awareness of environmental concerns, Commoner has been hailed by *Time* magazine as the "Paul Revere of Ecology" and by the *Earth Times* as "the dean of the environmental movement, who has influenced two generations." In 1980 Commoner was the presidential candidate of the Citizens Party, a liberal political party he helped found; the party's interests included an end to nuclear power, a switch to solar energy, and public control of the energy industry.

Commoner continues to lead the movement to eliminate pollution at its source. He asserts that prevention works while controls do not, and he publicly advocates the abandonment of fossil fuels as a primary source of energy. Having served society and the environment since the 1950's, he has succeeded in elevating the importance of considering environmental issues in all aspects of life.

Marcie L. Wingfield and Massimo D. Bezoari

FURTHER READING

Egan, Michael. *Barry Commoner and the Science of Survival: The Remaking of American Environmentalism.* Cambridge, Mass.: MIT Press, 2007.

Kriebel, David, ed. *Barry Commoner's Contribution to the Environmental Movement: Science and Social Action.* Amityville, N.Y.: Baywood, 2002.

Conservation policy

CATEGORY: Resources and resource management
DEFINITION: Laws and regulations designed to limit the economic exploitation of natural resources in the public interest
SIGNIFICANCE: Trends in conservation policy in the United States have varied in emphasis between allowing for limited exploitation of resources and prohibiting most economic uses of particular resources so that they can be preserved in an undeveloped state. Both approaches, however, have led to the protection and preservation of public lands and animal habitats.

The first use of the term "conservation" in relation to the natural environment was claimed by Progressive intellectuals in the early twentieth century. In his autobiography *Breaking New Ground* (1947), Gifford Pinchot, first chief of the U.S. Forest Service and a close friend of President Theodore Roosevelt, recalled realizing that all the natural resource problems are really one problem: the use of the earth for the permanent good of humans. The idea needed a name; presidential adviser Overton Price suggested "conservation," and the matter was settled.

ORIGINS OF CONSERVATION POLICY

The antecedents of modern conservation policies go back centuries. Aboriginal cultures around the world developed taboos governing behavior on the hunt. Venetians established reserves for deer and wild boar in the eighth century. Hunting reserves were common in Europe and Asia, and in colonial America those trees thought best for ships' masts were preserved for that purpose by decree.

Conservation policy, as the term is now understood, is a product of the nineteenth century, when population growth, urbanization, and industrialization created unprecedented opportunities for people to influence the natural world in which they lived, both for good and for ill. By the end of the nineteenth century, more Americans lived in cities than on farms. The nation was connected from coast to coast by telegraph and rail, and economic growth was rapid.

Most natural resource policies of the nineteenth century were designed to facilitate economic development. Best known among these policies was the Homestead Act of 1862, which gave free land to settlers, but similar policies provided free land to railroads and states. Other laws stimulated growth by providing for free use of timber and minerals on public lands.

The economic progress of the nineteenth century came at high cost to the environment, typified by the profligate use of forests and the near extermination of North American buffalo, and some prominent Americans took notice. In 1832 the artist and journalist George Catlin wrote of the probable extinction of the buffalo and American Indians, and he advocated a large national park where both might be preserved. Henry David Thoreau echoed Catlin's concerns in 1858, calling for national preserves. In 1864 George Perkins Marsh published *Man and Nature: Or, Physical Geography as Modified by Human Action*, the earliest important text with an ecological perspective.

As economic exploitation diminished the supply of natural resources, public attitudes began to change, and with them governmental policies. Although many laws continued to encourage economic growth, others reflected the growing desire for conservation. In 1864 the U.S. Congress sought to preserve the Yosemite Valley and Mariposa Big Tree Grove by giving them to the state of California for a public park. Eight years later Congress established the world's first national park at Yellowstone. In 1884 additional legislation prohibited all hunting and commercial fishing within Yellowstone National Park. In 1891 Congress established what would eventually become the national forest system when it authorized the president to set aside forest reserves on public lands.

PRESERVATION VERSUS WISE USE

These early conservation policies stressed resource preservation. Parks and forest reserves were simply set aside; none was effectively managed. The lack of management—especially forest management—displeased the advocates of scientific forestry, who also considered themselves conservationists. Early in its history, thus, the American conservation movement was divided, with some conservationists preaching "preservation" and others "wise use."

These contradictory tendencies were epitomized in the conflict between John Muir, founder of the Sierra Club, and Gifford Pinchot, principal architect and first chief of the U.S. Forest Service. Muir was the intellectual heir of Thoreau and Catlin. A perceptive scientist and popular author, he devoted his life to the exploration, enjoyment, and preservation of natural

(continued on page 34)

Milestones in Conservation Policy

Year	Event
1864	Yosemite Valley is ceded to California to create a park.
1872	The Yellowstone National Park Act establishes the world's first national park.
1891	The Forest Reserve Act authorizes the U.S. president to establish national forests.
1894	The Yellowstone Game Protection Act closes parks to hunting and commercial fishing.
1897	The Forest Management Act mandates that national forests be managed to perpetuate water supplies and wood products.
1900	The Lacey Act prohibits interstate shipment of wildlife that has been killed illegally.
1902	The Newlands Act establishes a national reclamation policy.
1903	The first National Wildlife Refuge is created at Pelican Island, Florida.
1905	The U.S. Forest Service is created within the Department of Agriculture to manage national forests.
1906	The Antiquities Act authorizes the creation of national monuments by presidential proclamation.
1910	The Pickett Act authorizes presidential land withdrawals for any public purpose.
1911	The Weeks Act provides for governmental purchase of national forestlands.
1913	The Hetch Hetchy Dam is authorized in Yosemite National Park.
1916	The National Park Service is created in the Interior Department to manage national parks.
1918	The Migratory Bird Treaty Act restricts the hunting of migratory birds.
1933	The Civilian Conservation Corps Act is passed.
1933	The Tennessee Valley Authority is created.
1934	The Taylor Grazing Act regulates grazing on public lands.
1937	The Federal Aid in Wildlife Restoration (Pittman-Robinson) Act provides federal aid to states for wildlife management.
1946	The U.S. Bureau of Land Management is created.
1950	The Federal Aid in Fish Restoration (Dingell-Johnson) Act provides federal aid to states for sport fish management.
1956	The Fish and Wildlife Act creates the U.S. Fish and Wildlife Service in the Interior Department.
1956	A proposal to construct Echo Park Dam in Dinosaur National Monument is defeated.
1960	The Multiple Use-Sustained Yield Act clarifies the purposes of national forests.
1964	The Wilderness Act establishes the National Wilderness Preservation System.
1964	The Land and Water Conservation Fund Act provides a trust fund for parkland acquisition.
1968	The National Wild and Scenic Rivers Act establishes a national river conservation system.
1968	The National Trails System Act establishes a national system of recreational trails.
1970	The National Environmental Policy Act requires environmental impact statements for federal activities that affect the environment.
1970	The Environmental Protection Agency (EPA) is created.
1970	Clean Air Act amendments establish stricter air-quality standards.
1971	The United Nations Educational, Scientific, and Cultural Organization (UNESCO) Biosphere Reserve Program recognizes areas of global environmental significance.
1972	The Clean Water Act establishes stricter water-quality standards.

Year	Event
1972	The United Nations Environmental Conference in Stockholm, Sweden, is attended by 113 nations.
1972	Federal Water Pollution Control Act amendments provide protection for wetlands.
1972	The Federal Environmental Pesticides Control Act requires pesticide registration.
1972	The Marine Mammal Protection Act imposes a moratorium on hunting or harassing of marine mammals.
1973	The Convention on International Trade in Endangered Species of Wild Fauna and Flora (CITES) prohibits international trade in endangered species.
1973	The Endangered Species Act commits the United States to the preservation of biological diversity.
1974	The Safe Drinking Water Act sets federal standards for public water supplies.
1976	The Toxic Substances Control Act authorizes the EPA to ban substances that threaten human health or the environment.
1976	The Federal Land Policy and Management Act directs the Bureau of Land Management to retain public lands and manage them for multiple uses.
1976	The Resource Conservation and Recovery Act directs the EPA to regulate waste production, storage, and transportation.
1976	The National Forest Management Act gives statutory protection to national forests and sets standards for management.
1977	The Surface Mining Control and Reclamation Act establishes environmental standards for strip mining.
1977	Clean Air Act amendments set high standards for air quality in large national parks and wilderness areas.
1980	The Fish and Wildlife Conservation Act provides federal aid for the protection of nongame wildlife.
1980	The Alaska National Interest Lands Conservation Act establishes more than 40.5 million hectares (100 million acres) of national parks and wildlife refuges in Alaska.
1980	The Comprehensive Environmental Response, Compensation, and Liability Act (CERCLA) establishes the Superfund hazardous waste cleanup program.
1982	The Nuclear Waste Policy Act establishes a process for siting a permanent nuclear waste repository.
1985	The U.S. government establishes the Conservation Reserve Program (CRP) to reduce agricultural surpluses by encouraging farmers to reduce the amounts of land they devote to crops, thereby helping to prevent soil erosion and reduce carbon in the atmosphere.
1987	The Montreal Protocol limits the production and consumption of chlorofluorocarbons (CFCs).
1988	The Ocean Dumping Act prohibits the dumping of sewage sludge and industrial waste.
1990	Clean Air Act amendments strengthen the Clean Air Act.
1992	The Earth Summit in Rio de Janeiro, Brazil, is attended by 179 nations.
1997	The Kyoto Protocol on Climate Change encourages global reduction in greenhouse gas emissions.
2001	The Roadless Area Conservation Rule, a federal policy initiative designed to protect national forests from commercial development, is issued.
2005	More than 4,000 hectares (approximately 10,000 acres) of Puerto Rican rain-forest land are added to the U.S. national forest system.
2008	The Food, Conservation, and Energy Act extends CRP enrollment authority through 2012.
2009	The Omnibus Public Land Management Act adds 850,000 hectares (2.1 million acres) of new wilderness areas in nine U.S. states.

ecosystems worldwide. Pinchot had studied scientific forestry at its source in Germany. He was a gifted politician, and his passion was not for preservation but for wise use. Muir believed that people could not improve on nature; his conservation was aesthetic and spiritual. Pinchot was committed to maximizing the human benefits from resource use through science; his conservation was economic and utilitarian. Although friends for a time, Muir and Pinchot eventually parted ways, with Muir becoming an advocate for preservation and national parks and Pinchot an advocate for wise use and national forests.

In the United States the legacy of Pinchot is alive and well in the multiple-use management principles of the Forest Service and the Bureau of Land Management and in organizations such as the Society of American Foresters, the International Society of Fish and Wildlife Managers, the National Rifle Association, and the Soil Conservation Society of America. Muir's emphasis on preservation has been institutionalized in the National Park Service and the Fish and Wildlife Service as well as in organizations such as the National Audubon Society, the Nature Conservancy, the Sierra Club, and the Wilderness Society.

Progressive Era

Many historians emphasize three eras of American conservation policy corresponding roughly to the Progressive Era, the New Deal, and the so-called environmental decade of the 1970's. The Progressive Era, epitomized by the presidency of Theodore Roosevelt, was the first golden age of American conservation policy. During this period Congress passed a number of pathbreaking conservation laws. The Lacey Act of 1900 put the power of federal enforcement behind state game laws, criminalizing the interstate transport of wildlife killed or captured in violation of state regulations. Another milestone of wildlife conservation was the ratification of a migratory bird treaty with Canada and passage of a law to enforce the treaty. With the Migratory Bird Treaty Act of 1918, the federal government asserted national authority to manage wildlife for conservation purposes, authority that was upheld by the U.S. Supreme Court in the case of *Missouri v. Holland* (1920).

Two critically important governmental agencies were created during this era: the Forest Service and the Park Service. In 1905 advocates of wise use and scientific forestry were rewarded with a Forest Service in the Department of Agriculture. The new agency's first director was Gifford Pinchot, the nation's foremost advocate of multiple-use forest management based on scientific principles. Under Pinchot's leadership the concepts of multiple use and sustained yield were applied in the rapidly growing national forest system. When Theodore Roosevelt became president, the United States had 18.6 million hectares (46 million acres) of national forest. By the end of his term of office, Roosevelt had increased the total size of the national forest system to 78.5 million hectares (194 million acres).

During the Progressive Era, advocates of preservation were often unsuccessful in their opposition to the policies of wise-use conservationists, but in the end they also had a victory. The most painful loss came in Yosemite National Park, where advocates of wise use joined forces with the city of San Francisco to dam the Hetch Hetchy Valley for a municipal water supply, forever destroying a natural valley some regarded as comparable to Yosemite Valley itself. The public outcry over the damming of Hetch Hetchy contributed to pressure for better park protection, however, and in 1916 preservationists achieved a long-sought goal: creation of the National Park Service to manage the growing system of fourteen national parks, including Yellowstone (1872), Yosemite and Sequoia (1890), Mount Rainier (1899), Crater Lake (1902), Wind Cave (1903), Mesa Verde (1906), Glacier (1910), Rocky Mountain (1915), and Hawaii and Lassen Volcanic (1916).

Two new forms of conservation reserves made their debut during this era: national wildlife refuges and national monuments. Roosevelt regarded wildlife sanctuaries as critical to the survival of game species. In 1903 he acted on his belief, creating the nation's first national wildlife refuge on Pelican Island in Florida. He had no specific legal authority to create a national wildlife refuge, but his usurpation was accepted at the time and later approved in principle. In 1910 the Pickett Act authorized the president to set aside land for any public purpose. National monuments began on a firmer foundation, but here too Roosevelt pushed conservation to the limit. The Antiquities Act of 1906 authorized the president to establish national monuments. As the name suggests, the law anticipated relatively small reservations to protect archaeological sites, but the monuments Roosevelt designated included 34,400 hectares (85,000 acres) at Petrified Forest, 120,600 hectares (298,000 acres) at Mount Olympus, and 326,200 hectares (806,000 acres) at the Grand Canyon. All of these later became national parks.

NEW DEAL ERA

The New Deal was, for the most part, a response to disaster. The primary disaster was the Great Depression, but the decade of the 1930's also saw the Dust Bowl, a minor climatic change that produced disastrous results on the Great Plains. New Deal conservation policies were responsive to the economic and ecological crises of the era, and they stressed wise use through scientific management rather than preservation. The Tennessee Valley Authority was created in 1933 to stimulate employment and economic growth in Appalachia through scientific management of the area's natural resources. The Taylor Grazing Act of 1934 was designed to end overgrazing of western public lands by imposing a system of permits based on principles of scientific management. The Civilian Conservation Corps and the Soil Conservation Service were both established during this era, and each contributed to repairing environmental damage. Greater concern for the management rather than the disposal of western public lands was also reflected in the creation in 1946 of the Bureau of Land Management to replace the General Land Office.

ENVIRONMENTAL DECADE

The so-called environmental decade lasted almost twenty years. It began with the inauguration of President John F. Kennedy, persisted through the presidential administrations of Lyndon B. Johnson, Richard Nixon, Gerald Ford, and Jimmy Carter, and ended with the inauguration of President Ronald Reagan. The conservation policies of this era were responsive to post-World War II economic growth that seemed to ensure economic prosperity while threatening quality of life. Stewart Udall, U.S. secretary of the interior, warned of a *Quiet Crisis* (1963), Rachel Carson of a *Silent Spring* (1962), Barry Commoner of *The Closing Circle* (1971), and Paul R. Ehrlich of *The Population Bomb* (1968). Conservation policy matured into environmental policy during this era. Conservation was still about husbanding natural resources, but to the historic concerns of conservation—such as forests, wilderness, and wildlife—were added concerns regarding clean water and clean air, energy supplies, and the problems posed by hazardous and toxic wastes. During this era Congress passed most of the major laws that continue to shape conservation policy at the beginning of the twenty-first century.

Preservation policy was strengthened. In 1964 Congress passed the Wilderness Act and the Land and Water Conservation Fund Act. The former established the National Wilderness Preservation System, which has grown from more than 3.6 million hectares (9 million acres) to more than 40.5 million hectares (100 million acres). The latter facilitated acquisition of land for parks and open space. Four years later Congress established a national system of trails as well as a national system to protect wild and scenic rivers from certain kinds of development. Both the Forest Service and the Bureau of Land Management were given new statutory direction emphasizing planning and preservation, sometimes at the expense of economic development. At the end of the era Congress passed the Alaska National Interest Lands Conservation Act (1980), making conservation withdrawals of more than 40.5 million hectares of public lands and doubling the size of the national park and wildlife refuge systems nationwide.

Wise-use conservation was also well served as Congress radically increased federal regulation of resource use. President Johnson established a presidential commission on natural beauty and addressed world population and resource scarcity in his 1965 state of the union speech. Environmental management was nationalized through a series of far-reaching statutes addressing air pollution, water pollution, marine resources, noise pollution, biological diversity, toxic chemicals, and hazardous waste. New burdens were placed on government and private citizens. The National Environmental Policy Act of 1969 required all governmental agencies to study the probable environmental effects of their actions before moving forward. A large number of environmental enforcement programs were reorganized in 1970 into the newly created Environmental Protection Agency (EPA). The EPA is an independent agency, but presidents have routinely regarded its director as having cabinet status.

CONSERVATION POLICY FOR A NEW MILLENNIUM

At the policy level, the era following the environmental decade was one of consolidation rather than new initiatives. The Reagan administration was hostile to environmental policy and attempted to tilt public policy toward less environmental regulation. President Reagan was able to prevent the adoption of major new conservation policies, but his administrative efforts—led by Interior Secretary James Watt—to roll back environmental laws were successfully resisted by Congress. A major new air-pollution statute was passed during the presidential administration of George H. W. Bush, but this was exceptional for the era. Bill Clinton's presidential administration gave greater attention to conservation policy, but in the years following the 1994 elections there was little cooperation between the president and Congress on environmental issues.

Americans have come to expect government to practice conservation and protect environmental quality. Doing so is increasingly difficult. Beyond the policy gridlock of the 1980's and 1990's, the issues themselves have become more difficult. The most pressing issues—such as stratospheric ozone depletion, climate change, and biological diversity—are beset by scientific uncertainty. They are also global in scope and thus beyond the ability of any single nation to address independently. The future of conservation policy appears to be in the international arena, where extant institutions lack the authority to govern. Treaties addressing the use of chlorinated fluorocarbons, the preservation of biological diversity, and the limitation of greenhouse gas emissions demonstrate that nations are giving increasing attention to conservation issues, but international achievements remain modest.

Craig W. Allin

FURTHER READING

Allin, Craig W. *The Politics of Wilderness Preservation*. 1982. Reprint. Fairbanks: University of Alaska Press, 2008.

Chiras, Daniel D., and John P. Reganold. *Natural Resource Conservation: Management for a Sustainable Future*. 10th ed. Upper Saddle River, N.J.: Benjamin Cummings/Pearson, 2010.

Davis, David Howard. *American Environmental Politics*. Belmont, Calif.: Wadsworth, 1998.

Dowie, Mark. *Conservation Refugees: The Hundred-Year Conflict Between Global Conservation and Native Peoples*. Cambridge, Mass.: MIT Press, 2009.

French, Hilary. *Vanishing Borders: Protecting the Planet in the Age of Globalization*. New York: W. W. Norton, 2000.

Hayes, Samuel P. *Conservation and the Gospel of Efficiency: The Progressive Conservation Movement, 1890-1920*. 1959. Reprint. Pittsburgh: University of Pittsburgh Press, 1999.

Nash, Roderick. *Wilderness and the American Mind*. 4th ed. New Haven, Conn.: Yale University Press, 2001.

Rosenbaum, Walter A. *Environmental Politics and Policy*. 7th ed. Washington, D.C.: CQ Press, 2008.

Convention on International Trade in Endangered Species

CATEGORIES: Treaties, laws, and court cases; animals and endangered species

THE CONVENTION: International agreement aimed at conserving endangered animal and plant species

DATE: Opened for signature on March 3, 1973

SIGNIFICANCE: The Convention on International Trade in Endangered Species was the first international agreement concerning the conservation of wildlife that constituted a legal commitment by the parties to the convention and also included a means of enforcing its provisions.

Until the 1970's, international agreements that had been made to address the preservation of species did not include any binding legal commitment on the part of the countries signing them; thus they were ineffectual in protecting the species they were written to protect. In 1969, however, the United States passed the Endangered Species Conservation Act, which contained a provision that gave the secretaries of interior and commerce until June 30, 1971, to call for an international conference on endangered species. The resulting conference, which was held in Washington, D.C., in March, 1973, resulted in the Convention on International Trade in Endangered Species of Wild Fauna and Flora (CITES).

The United States was the first country to ratify the convention, which entered into effect on July 1, 1975; by 2010, 175 nations in total were parties to the agreement. CITES is intended to conserve species and does this by managing international trade in those species. It was the first international convention on the conservation of wildlife that constituted a legal commitment by the parties to the convention and also included a means of enforcing its provisions. This enforcement includes a system of trade sanctions and an international reporting network to stop trade in endangered species. The system established by CITES does, however, contain loopholes through which states with special interest in particular species can opt out of the global control for those species.

A major feature of CITES is its categorization of three levels of vulnerability of species. Appendix I includes all species that are threatened with extinction and whose status may be affected by international trade. Appendix II includes species that are not yet threatened but might become endangered if trade in them is not regulated. It also includes other species that, if traded, might affect the vulnerability of the first group. Appendix III lists species that a signatory party identifies as subject to regulation in order to restrict exploitation of that species. The parties to the treaty agree not to allow any trade in the species on the three lists unless an exception is allowed in CITES.

The species listed in the appendixes may be moved from one list to another as their vulnerability increases or decreases. According to the convention, states may implement stricter measures of conservation than those specified in the convention or may ban trade in species not included in the appendixes. CITES also establishes a series of import and export trade permits within each of the categories. Each nation designates a management authority and a scientific authority to implement CITES. Exceptions to the ban on trade are made for scientific and museum specimens, exhibitions, and movement of species under permit by a national management authority.

The parties to CITES maintain records of trade in specimens of species that are listed in the appendixes and prepare periodic reports on their compliance with the convention. These reports are sent to the CITES secretariat in Switzerland, administered by the United Nations Environment Programme (UNEP), which issues notifications to all parties of state actions and bans. The secretariat's functions are established by the convention and include interpreting the provisions of CITES and advising countries on implementing those provisions by providing assistance in writing their national legislation and organizing training seminars. The secretariat also studies the status of species being traded in order to ensure that the exploitation of such species is within sustainable limits.

The CITES Conference of Parties meets every two or three years in order to review implementation of the convention. The meetings are also attended by nonparty states, intergovernmental agencies of the United Nations, and nongovernmental organizations considered "technically qualified in protection, conservation or management of wild fauna and flora."

The meetings are held in different signatories' countries: The first took place in Berne, Switzerland, on November 2-6, 1976. At the conference, the parties may adopt amendments to the convention and make recommendations to improve the effectiveness of CITES.

CITES has been incorporated into Caring for the Earth: A Strategy for Sustainable Living. This strategy was launched in 1991 by UNEP, the International Union for Conservation of Nature (IUCN), and the World Wide Fund for Nature (WWF). Other nongovernmental groups working to support CITES are Fauna and Flora International (FFI), Trade Records Analysis of Flora and Fauna in Commerce (TRAFFIC International), and the World Conservation Monitoring Centre (WCMC).

Some of the species protected by CITES have received additional protection under later agreements. In certain cases, however, states have allowed trade in listed species to continue for economic purposes or have refused to sign CITES because of the extent to which they trade in a species or species part, such as ivory. Others have signed because they needed help in stopping illegal trade and poaching of species within their borders. Whales have proven to be a difficult species to protect. Whales are given protection under CITES according to the status of specific species. The moratorium on commercial whaling instituted in 1986 by the International Whaling Commission (IWC) was intended to strengthen the CITES protection by species, but the whaling states have disagreed on the numbers of whale populations, and some have withdrawn from the IWC and resumed their whaling activities.

Colleen M. Driscoll

Species Protected by CITES

Species Type	Species	Subspecies	Populations
Mammals	617	36	26
Birds	1,455	17	3
Reptiles	657	9	10
Amphibians	114	0	0
Fish	86	0	0
Invertebrates	2,179	5	0
Total fauna	5,108	67	39
Plants	28,977	7	3
Total	34,085	74	42

FURTHER READING

Chasek, Pamela S., et al. *Global Environmental Politics*. 4th ed. Boulder, Colo.: Westview Press, 2006.

Hutton, Jon, and Barnabas Dickson, eds. *Endangered Species, Threatened Convention: The Past, Present, and Future of CITES*. Sterling, Va.: Earthscan, 2000.

International Union for Conservation of Nature. *Conserving the World's Biological Diversity*. Washington, D.C.: Island Press, 1990.

Van Dyke, Fred. "The Legal Foundations of Conservation Biology." In *Conservation Biology: Foundations, Concepts, Applications*. 2d ed. Ne2w York: Springer, 2008.

Cousteau, Jacques

CATEGORIES: Activism and advocacy; animals and endangered species
IDENTIFICATION: French explorer, conservationist, and filmmaker
BORN: June 11, 1910; Saint-André-de-Cubzac, France
DIED: June 25, 1997; Paris, France
SIGNIFICANCE: Cousteau, one of the twentieth century's best-known explorers and conservationists, gained widespread attention for environmental issues, particularly those concerning the world's oceans.

Jacques Cousteau was, in his own mind, neither scientist nor adventurer. Rather, he considered himself a filmmaker. In 1971 Cousteau said, "I am not a scientist, I am rather an impresario of scientists." Cousteau, however, probably did recognize the vital role he played in exciting public interest in the ocean and revealing the intricacies of humanity's relationship to it.

Cousteau's long career began in 1933 with a stint in the French navy. In 1943, during his military service, he developed, with Émile Gagnan, the Aqua-Lung, the first commercially available self-contained underwater breathing apparatus (scuba). In 1950 Cousteau bought the retired minesweeper *Calypso* for use as a research ship. The many films and books Cousteau authored in his lifetime helped finance the *Calypso*'s expeditions, which brought the wonders of the oceans to an international audience. He is perhaps best known to many Americans from his documentary television series *The Undersea World of Jacques Cousteau*, which aired from 1966 to 1976.

By 1956 Cousteau's research activities had become a full-time career, so he resigned from the navy. In 1957 Prince Rainier III named Cousteau director of the Oceanographic Museum of Monaco, a post he held for thirty-one years. During the 1950's and 1960's Cousteau sometimes explored the same sites several times. As a result of these trips, he eventually recognized that human activities were degrading the aquatic environment.

In 1958 Cousteau helped establish the world's first undersea marine reserve off the coast of Monaco. In 1960 he spoke out against France's plans to dispose of nuclear waste in the Mediterranean Sea. One decade later, he summed up his fears about ocean pollution by stating, "The oceans are in danger of dying."

In 1974 Cousteau founded the Cousteau Society, a nonprofit organization "dedicated to the protection and improvement of the quality of life for present and future generations." Among other achievements, this group has provided logistical support and facilities for hundreds of scientists, helped develop postgraduate

Explorer and conservationist Jacques Cousteau. (Library of Congress)

environmental science and policy programs at universities worldwide, and worked toward the adoption by the United Nations of a "Bill of Rights for Future Generations"—all causes to which Cousteau himself was strongly committed. In addition to revealing the ocean's majesty, Cousteau strove to alert the public to the dilemma he recognized during his explorations:

> The pursuit of technology and progress may today endanger the very survival of . . . practically all life on earth. . . . [However] the technology that we use to abuse the planet is the same technology that can help us to heal it.

Cousteau received numerous awards and honors for his environmental efforts, including the United Nations International Environmental Prize (1977) and a place on the United Nations Global 500 Roll of Honor (1988). He served as adviser for environmental sustainable development to the World Bank and was a member of the United Nations High-Level Board on Sustainable Development (1992). In 1992 he was appointed chairman of the newly created Council on the Rights of Future Generations. Three years later he resigned from the council in protest over the resumption of nuclear weapons testing in the South Pacific.

Clayton D. Harris

Further Reading

Kroll, Gary. *America's Ocean Wilderness: A Cultural History of Twentieth-Century Exploration.* Lawrence: University Press of Kansas, 2008.

Matsen, Brad. *Jacques Cousteau: The Sea King.* New York: Pantheon Books, 2009.

Darling, Jay

Categories: Activism and advocacy; animals and endangered species
Identification: American cartoonist and wildlife conservationist
Born: October 21, 1876; Norwood, Michigan
Died: February 12, 1962; Des Moines, Iowa
Significance: Darling was instrumental in starting the federal duck stamp program, which generated revenue to buy new lands to serve as waterfowl refuges, as authorized by the Migratory Bird Conservation Act.

Jay Darling was an editorial cartoonist by profession. He was known among his peers as Ding, the name with which he signed his drawings. His biting cartoons, which won him two Pulitzer Prizes and national recognition, often depicted the destruction of America's waterfowl and their habitat by overhunting and periodic drought—particularly during the Dust Bowl of the 1930's, which dried the wetlands that the birds required.

The passage of the Migratory Bird Conservation Act by the U.S. Congress in 1929 laid the groundwork for Darling's major contribution to waterfowl conservation. That act authorized the U.S. Department of Agriculture to acquire wetlands and preserve them as waterfowl habitat, but it provided no permanent source of funding for the purpose. In 1934 President Franklin D. Roosevelt appointed a committee to look into the need for waterfowl refuges. Among its members were Darling, who had been a fierce critic of Roosevelt's wildlife policies, wildlife conservationist Aldo Leopold, and publisher Thomas Beck. The committee's recommendation that $50 million be spent for new refuges rekindled an idea that had lain dormant for years: to require waterfowl hunters to pay fees for the privilege of hunting each year by buying stamps known as duck stamps. The revenue generated from the stamps would be used to buy new refuge lands authorized by the Migratory Bird Conservation Act.

In March, 1934, Congress passed the Migratory Bird Hunting and Conservation Stamp Act, which required waterfowl hunters sixteen years or older to buy an annual duck stamp. That same month, Roosevelt appointed Darling chief of the Department of Agriculture's Bureau of Biological Survey, forerunner of the U.S. Fish and Wildlife Service. While serving as chief, Darling carried out the 1934 act's mandate by initiating the federal duck stamp program; Darling also designed the first duck stamp.

It took some time for money from the sale of the duck stamps to start flowing, however, so Darling began raising money from other programs within the Department of Agriculture to purchase refuge land. He is credited with obtaining $20 million for wildlife conservation and setting aside 1.8 million hectares (4.5 million acres) as refuge land during his twenty-month tenure at the Bureau of Biological Survey. He resigned in November, 1935, dismayed by conservationists' lack of collective strength in focusing attention on wildlife issues.

At the first government-sponsored North American Wildlife Conference in 1936, Darling helped to organize the 36,000 wildlife societies then in existence across the United States into a national body called the General Wildlife Federation. Darling was unanimously chosen president of the federation. Two years later, the group's name was changed to the National Wildlife Federation, and Darling was reelected its president.

In 1942 Darling was awarded the Theodore Roosevelt Medal for distinguished service in wildlife conservation; in 1960 he received a National Audubon Society medal for distinguished service in natural resource conservation. A 2,013-hectare (4,975-acre) wildlife refuge established on Sanibel Island, Florida, in 1945 was dedicated to Darling in 1978. Now known as the J. N. "Ding" Darling National Wildlife Refuge, it supports a wide diversity of birds and other animals.

Jane F. Hill

FURTHER READING

Darling, Jay N. *J. N. "Ding" Darling's Conservation and Wildlife Cartoons.* Des Moines, Iowa: J. N. "Ding" Darling Foundation, 1991.

Jonson, Laurence F. *Federal Duck Stamp Story: Fifty Years of Excellence.* Davenport, Iowa: Alexander, 1984.

Lendt, David L. *Ding: The Life of Jay Norwood Darling.* 4th ed. Mt. Pleasant, S.C.: Maecenas Press, 2001.

The first Duck Stamp, featuring art by Jay Darling. (USFWS)

Earth First!

CATEGORIES: Organizations and agencies; activism and advocacy; preservation and wilderness issues
IDENTIFICATION: Radical environmental movement
DATE: Founded in 1980
SIGNIFICANCE: Earth First! gained notoriety during the 1980's for employing unorthodox and controversial means to protest the abuse of wilderness areas. The movement's use of ecologically motivated sabotage and highly visible civil disobedience actions, such as tree sitting to prevent logging, has drawn public attention to environmental causes at the same time it has drawn criticism from some environmentalists.

Earth First! was founded in the United States in 1980 by five men who were concerned about what they perceived to be a lack of passion and commitment within the old-line conservation organizations. The five founders were Howie Wolke, a Friends of the Earth representative; Ron Kezar, a National Park Service worker; Bart Koehler, a Wilderness Society representative; Mike Roselle, a Yippie activist; and Dave Foreman, a lobbyist for the Wilderness Society who had once been a Barry Goldwater Republican. Other like-minded persons quickly joined the cause.

A movement rather than a formally structured organization, Earth First! has gained notoriety and sparked controversy through its environmental activism. The movement is committed to naturalist John Muir's wilderness preservation ideals as well as to the principles of deep ecology. The philosophy of Earth First! is expressed in the slogan "No compromise in defense of Mother Earth." Members of the movement view private and public developers as their enemies, and they see the U.S. Forest Service as being too willing to capitulate to developers. The movement is also highly critical of mainstream conservation organizations, which Earth First! accuses of having become passive and bureaucratized.

Earth First!ers use various strategies to inhibit the destruction of wilderness, from letter-writing campaigns, petitions, boycotts, and legal actions to demonstrations, nonviolent civil disobedience, and ecologically motivated sabotage (also called ecotage or monkeywrenching). The movement entered the media spotlight in 1981 when Earth First!ers unfurled a 300-foot-long piece of black polyethylene depicting a "crack" on the face of the Glen Canyon Dam, located just south of the Utah-Arizona border, in symbolic opposition to the dam and its recently created reservoir, Lake Powell. Their methods led the media to dub Earth First!ers a real-life Monkey Wrench Gang, for the titular heroes of Edward Abbey's 1975 novel who sabotage construction equipment in the American Southwest to stop environmental destruction.

Black plastic quickly gave way to monkeywrenching: Earth First!ers pulled up survey stakes, blocked logging roads, disabled construction equipment, and spiked old-growth trees (a controversial tactic in which metal or ceramic spikes are pounded into trees to ruin their commercial use and thwart logging equipment). Such actions have been widely condemned and criticized. Tree spiking, in particular, has the potential to harm loggers and sawmill workers by shattering saw blades that contact spikes; it was made illegal in the United States in 1988. Foreman published a detailed how-to book on ecotage, *Ecodefense: A Field Guide to Monkeywrenching* (1985), in which he—like the movement itself—claimed to neither condemn nor condone the use of such tactics.

In 1989, Foreman and three Arizona Earth First!ers were arrested in an undercover sting operation conducted by the Federal Bureau of Investigation (FBI); they were charged with planning to sabotage power lines. Foreman was released on probation, but the others served prison sentences—a fact that caused resentment against Foreman among some Earth First!ers. In the decade since the movement's inception, a rift had developed within the movement between conservationists such as Foreman and social justice activists such as Roselle. Earth First!'s anti-hierarchical lack of structure and leadership had attracted a politically anarchic following, and the movement's focus had broadened from wilderness preservation to a host of leftist causes. In 1990 Foreman and others dissatisfied with what Earth First! had become left the movement.

That year marked Earth First!'s Redwood Summer—a conscious attempt to emulate the Civil Rights movement's 1964 Freedom Summer—when a number of protesters faced off against the lumber industry in California's redwood forests. The protesters demonstrated a more communal, countercultural, and leftist spirit, with greater involvement by women, an increasing emphasis on social justice, and attempts to appeal to the loggers as a class against the corporations. One of the organizers, Judi Bari, was seriously injured by a bomb that had been placed in her automobile by an unknown assailant. Earth First! contin-

---

Ethics of an Eco-Warrior

In the following excerpt from his memoir Confessions of an Eco-Warrior *(1991), environmentalist Dave Foreman conveys the sense of passion and frustration that ultimately led to his founding of the grassroots organization Earth First!*

We, as human beings, as members of industrial civilization, have no divine mandate to pave, conquer, control, develop, or use every square inch of this planet. As Edward Abbey . . . said, we have a right to be here, yes, but not everywhere, all at once.

The preservation of wilderness is not simply a question of balancing competing special-interest groups, arriving at a proper mix of uses on our public lands, and resolving conflicts between different outdoor recreation preferences. It is an ethical and moral matter. A religious mandate. Human beings have stepped beyond the bounds; we are destroying the very process of life. . . .

The crisis we now face calls for *passion*. When I worked as a conservation lobbyist in Washington, D.C., I was told to put my heart in a safe deposit box and replace my brain with a pocket calculator. I was told to be rational, not emotional, to use facts and figures, to quote economists and scientists. I would lose credibility, I was told, if I let my emotions show.

But, damn it, I am an animal. A living being of flesh and blood, storm and fury. The oceans of the Earth course through my veins, the winds of the sky fill my lungs, the very bedrock of the planet makes my bones. I am alive! I am not a machine, a mindless automaton, a cog in the industrial world, some New Age android. When a chain saw slices into the heartwood of a two-thousand-year-old Coast Redwood, it's slicing into my guts. When a bulldozer rips through the Amazon rain forest, it's ripping into my side. When a Japanese whaler fires an exploding harpoon into a great whale, my heart is blown to smithereens. I am the land, the land is me.

ued its confrontational campaign to save the redwoods through the decade, with Julia "Butterfly" Hill's two-year-long occupation of an old-growth redwood tree in 1997-1999 gaining international media attention. In 1998 fellow redwood-defending Earth First!er David "Gypsy" Chain lost his life when he was struck by a tree that a logger refused to stop cutting.

Earth First! groups sprang up in the United Kingdom during the early 1990's, as did a radical spin-off group called the Earth Liberation Front (ELF). As Earth First! began to work to distance itself from criminal activities, focusing instead on nonviolent civil disobedience, ELF embraced economic sabotage as a means to save a dying planet. ELF, which went on to become an international presence, regards extreme methods such as arson and bombing as legitimate weapons in the war against profit-motivated environmental destruction. Housing developments, chain stores, construction sites, and automotive dealerships selling sport utility vehicles are favorite ELF targets. In 2002 the FBI listed ELF as America's largest and most active domestic terrorist group. Spokespersons for ELF maintain that the group advocates violent action only against inanimate objects, but it does not shy away from harassment and threats against developers and corporate heads—"the real terrorists," according to ELF.

By 2010 the Web site of the *Earth First! Journal* listed Earth First! groups in Canada, Iceland, the United Kingdom, Ireland, Belgium, the Netherlands, Germany, Italy, the Czech Republic, Poland, Russia, Nigeria, the Philippines, and Australia, in addition to groups in twenty-seven U.S. states.

Eugene Larson
Updated by Karen N. Kähler

FURTHER READING

Foreman, Dave, and Bill Haywood, eds. *Ecodefense: A Field Guide to Monkeywrenching*. 3d ed. Chico, Calif.: Abbzug Press, 2002.

Liddick, Donald R. *Eco-Terrorism: Radical Environmental and Animal Liberation Movements*. Westport, Conn.: Praeger, 2006.

Scarce, Rik. *Eco-Warriors: Understanding the Radical Environmental Movement*. Updated ed. Walnut Creek, Calif.: Left Coast Press, 2006.

Wall, Derek. *Earth First! and the Anti-Roads Movement: Radical Environmentalism and Comparative Social Movements*. New York: Routledge, 2002.

Echo Park Dam opposition

CATEGORIES: Activism and advocacy; preservation and wilderness issues
THE EVENT: Environmentalists' efforts to halt the building of a dam in the Echo Park river bottom in Utah's Dinosaur National Monument
DATES: 1952-1956
SIGNIFICANCE: The Sierra Club's successful attempt to stop construction of the proposed Echo Park Dam in Dinosaur National Monument during the 1950's signaled the emergence of the environmental movement as a powerful political force in the United States.

The U.S. Bureau of Reclamation first suggested building a high dam at the Echo Park site, located 3.2 kilometers (2 miles) below the confluence of the Green and Yampa rivers on the border of Utah and Colorado, during the 1930's. No formal request was made to the U.S. Congress for authorization of the project until 1950, however. The terms of the Organic Act of 1916, passed following the controversy surrounding construction of the Hetch Hetchy Dam in Yosemite National Park in California, prohibited such a project, but administrators within the bureau believed legislators would be willing to make an exception for Echo Park. Subsequent events proved them wrong.

At the time that the Bureau of Reclamation asked Congress for permission to build a high dam within the boundaries of Dinosaur National Monument, few people expected any significant opposition. In the years following the Great Depression, both the U.S. government and the general public saw dam development as good for the economy and thus good for the country. True, noted photographer Ansel Adams had helped mobilize opposition to hydroelectric development on the Kings River in California a decade earlier, leading to the creation of Kings Canyon National Park, but the Kings River area was home to giant sequoia trees—its scenic wilderness value was obvious. Dinosaur National Monument, in contrast, appeared barren. As long as the Echo Park Dam would not inundate the dinosaur fossil quarries, advocates of wilderness preservation initially voiced few objections. According to David Brower, executive director of the Sierra Club at the time, even members of his organization described the monument as being nothing but sagebrush.

This changed in 1952 following a Sierra Club member's trip through Dinosaur National Monument. The home movie footage he shot of the canyons within the monument persuaded Brower and others to take a closer look. In 1953 the Sierra Club began organizing rafting trips along the Green River through Dinosaur National Monument. As more people traveled through the spectacular river canyons, opposition to dam construction within the boundaries of the monument grew. Other wilderness preservation and conservation groups, such as the Wilderness Society and the Izaak Walton League, along with prominent writers and politicians, joined with the Sierra Club in fighting the Echo Park Dam proposal.

In 1956 this coalition of preservationists and conservationists was successful: The Bureau of Reclamation dropped its plans for the Echo Park Dam. The victory for the environmental preservationists proved bittersweet, however. In exchange for the cancellation of the plans for Echo Park, the environmentalists agreed not to fight the Bureau of Reclamation's plan to build Glen Canyon Dam on the Colorado River, a decision Brower later regretted. Still, by preventing construction of the Echo Park Dam, the environmentalists reaffirmed the important principle that no industrial development should ever take place within a national park.

Nancy Farm Männikkö

FURTHER READING

Lowry, William Robert. *Dam Politics: Restoring America's Rivers.* Washington, D.C.: Georgetown University Press, 2003.

Palmer, Tim. "The Beginnings of River Protection." In *Endangered Rivers and the Conservation Movement.* 2d ed. Lanham, Md.: Rowman & Littlefield, 2004.

Ecotage

CATEGORIES: Activism and advocacy; philosophy and ethics

DEFINITION: Sabotage tactics used by radical environmentalists to stop projects they perceive as ecologically destructive

SIGNIFICANCE: Many mainstream environmentalists believe that the actions of those who engage in so-called ecotage have alienated some people who would otherwise support the environmentalist cause, but others assert that such extreme acts are sometimes necessary to call attention to the need to defend nature.

In 1972 the group Environmental Action published the handbook *Ecotage!*, which compiled ideas about how to sabotage environmentally destructive projects. Edward Abbey's novel *The Monkey Wrench Gang* (1975), featuring a small group of ecoguerrillas who destroy construction equipment to stop development in the southwestern United States, inspired Dave Foreman and others to start the radical environmental movement Earth First! In 1985 Foreman published *Ecodefense: A Field Guide to Monkeywrenching*, a manual of ecotage methods and information on related issues such as safety and security. In the early 1990's Foreman wrote that "those willing to commit ecotage are needed today as never before." Several other environmental advocates, such as Greenwar International, have also promoted ecotage.

The early proponents of ecotage, known as ecoteurs, were dismayed by industrial development of wilderness areas that the government refused to protect. Frustrated that civil disobedience did not achieve their goals, ecoteurs decided to preserve the environment by illegally damaging machinery used to degrade wilderness areas. Such militant acts of destruction often impeded future development efforts and reduced industrialists' profits.

Many ecoteurs, including Foreman, had belonged to mainstream environmental groups during the 1960's and 1970's but had become disillusioned by the dominance of conservative political leaders in these groups during the 1980's. Ecoteurs are often critical of environmentalists in such organizations as the Sierra Club, asserting that they are passive, ignore opportunities to preserve the wilderness, and appease industrialists and governmental agencies by sacrificing the environment. Ecoteurs denounce environmentalists with anthropocentric attitudes who view the environment as a source of production and resources to fulfill human needs.

Most ecoteurs consider themselves a symbiotic part of the environment and justify their bold, destructive conduct as acts of self-defense. "It is time to act heroically and admittedly illegally in defense of the wild," Foreman stated, "to put a monkeywrench into the gears of the machinery destroying natural diversity." He stressed that ecotage "is sabotage, not terrorism, because it's about property destruction. It's saying,

Ecoterrorism

CATEGORIES: Activism and advocacy; philosophy and ethics

DEFINITION: Clandestine activities conducted by radical environmentalists with the intent of disrupting environmental damage or preventing cruelty to animals

SIGNIFICANCE: The extreme activities of some radical environmentalists have caused considerable controversy, with critics noting that ecoterrorists have damaged the reputations of many hardworking environmentalist groups around the world.

Ecoterrorism began in the United States in the 1970's, peaked in the early 1990's, and waned later in that decade, although acts of ecoterrorism continued into the twenty-first century. Other terms, such as "ecological sabotage," "monkeywrenching," "ecotage," and "decommissioning," roughly connote the same concept, but with much less pejorative implications. "Monkeywrenching" was coined by Edward Abbey in his novel *The Monkey Wrench Gang* (1975). "Ecotage" refers to acts of sabotage for environmental ends. The radical environmentalist group Earth First! probably engaged in more ecoterrorist activities during the peak period than any other group, including People for the Ethical Treatment of Animals (PETA), the Sea Shepherd Conservation Society, and the Animal Liberation Front (ALF). In 1985 Dave Foreman, cofounder of Earth First!, published *Ecodefense: A Field Guide to Monkeywrenching*, which describes a variety of radical environmental activities and ways to carry them out.

Lorenz Otto Lutherer and Margaret Sheffield Simon chronicle the history of ecoterrorism conducted by animal rights groups in *Targeted: The Anatomy of an Animal Rights Attack* (1992). They describe numerous cases of break-ins, vandalism, thefts of animals and equipment, and threats of violence against researchers and businesspeople. Animal rights activists who engage in these activities, these authors maintain, are a menace to society, scientific and economic progress, and democracy itself.

The types of activities that initially led critics to characterize particular environmental activists as ecoterrorists can be divided into three categories. The first involves legal protests that are still clear cases of monkeywrenching. Examples are sit-ins in front of

I'm operating as part of the wilderness, defending myself."

Sometimes calling themselves "ecowarriors," ecoteurs strategically select their targets for ecotage in their efforts to disrupt environmentally harmful development. Many have focused on obstructing logging and strip-mining operations. Common monkeywrenching procedures include tree spiking, in which large nails are driven into trees in old-growth forests to deter loggers. Other tactics include pouring sand into the gas tanks of trucks and bulldozers, as well as slashing tires. Ecoteurs have also pulled up survey stakes, cut power lines, blockaded roads, stolen machinery parts, and ruined tools. Some ecoteurs have damaged offices belonging to the U.S. Forest Service to protest logging in national forests, and others have sprayed baby seals with dye so their fur would be unusable for coats. The Federal Bureau of Investigation (FBI) and industrial leaders have offered rewards for the arrest and conviction of ecoteurs.

During the 1990's, many of the people in the Earth First! movement came to believe that ecotage often does more to ostracize people than to protect the environment, and they reassessed their protest tactics. Columns in the *Earth First! Journal* during that period discussed individuals' ecotage efforts, but Earth First!ers disagreed about the role of ecotage in the environmental movement. While some believed all ecotage should be ceased, others argued that only certain acts of sabotage should be curtailed—those that might result in injuries to loggers or miners. Although some ecoteurs began to question their tactics in the 1990's, ecotage continued into the twenty-first century.

Elizabeth D. Schafer

FURTHER READING

Foreman, Dave, and Bill Haywood, eds. *Ecodefense: A Field Guide to Monkeywrenching*. 3d ed. Chico, Calif.: Abbzug Press, 2002.

Scarce, Rik. *Eco-Warriors: Understanding the Radical Environmental Movement*. Updated ed. Walnut Creek, Calif.: Left Coast Press, 2006.

Sterba, James P., ed. *Earth Ethics: Introductory Readings on Animal Rights and Environmental Ethics*. 2d ed. Upper Saddle River, N.J.: Prentice Hall, 2000.

Time Line of Ecoterror Incidents

Year	Incident
1998	Arson of a U.S. Department of Agriculture animal damage control building near Olympia, Washington, resulting in $2 million in damages
1998	Arson of a Vail, Colorado, ski resort, resulting in $12 million in damages
1999	Arson at a genetic-engineering research office at Michigan State University
2003	Arson at a HUMMER dealership in West Covina, California, destroying 125 SUVs and resulting in an estimated $1 million in damages
2005	Arson of five townhouses under construction in Hagerstown, Maryland
2008	Arson at the Street of Dreams housing development in Woodinville, Washington, resulting in $12 million in damages

offices, laboratories, factories, bulldozers, and even locomotives, sometimes accentuated by activists' chaining themselves to gates or trees. Activists have also often engaged in a process called tree spiking, in which metal spikes are driven into trees to prevent logging with chain saws.

The activities in the second category are characterized by their illegal nature, such as decommissioning machinery by pouring sand, sugar, or water into gas tanks and damaging motor vehicles by smashing distributors and spark plugs. Most break-ins at animal research facilities are included in this category. The third category consists of more daring but potentially hazardous operations, such as the ramming of whaling ships by Sea Shepherd activists.

Although many environmentalists oppose acts of violence in the name of their cause, others view monkeywrenching activists as "the conscience of the environmental movement." However, most agree that the term "ecoterrorism" more aptly applies to those who plunder the earth and its atmosphere in the name of capitalism and progress. Some environmental activists compare themselves to Resistance fighters of World War II; they see environmental damage as equivalent to the Holocaust and other war crimes, meriting drastic reprisals. In general, however, environmental activists believe that their actions ought to be nonviolent, and most who participate in ecotage are aggressive in taking preventive measures. For example, to avoid injuries, antilogging activists inform loggers and mill workers about trees that contain spikes by sending letters, telephoning, or marking trees with paint.

Opinions regarding the moral implications of ecoterrorism vary. Many environmental activists, whether or not they approve, agree that it works. However, critics and even some activists feel that some radical activists have damaged the reputations of many hardworking, peace-loving environmentalists around the world.

Chogollah Maroufi

FURTHER READING

Foreman, Dave, and Bill Haywood, eds. *Ecodefense: A Field Guide to Monkeywrenching*. 3d ed. Chico, Calif.: Abbzug Press, 2002.

Lutherer, Lorenz Otto, and Margaret Sheffield Simon. *Targeted: The Anatomy of an Animal Rights Attack*. Norman: University of Oklahoma Press, 1992.

Scarce, Rik. *Eco-Warriors: Understanding the Radical Environmental Movement*. Updated ed. Walnut Creek, Calif.: Left Coast Press, 2006.

Ehrlich, Paul R.

CATEGORIES: Activism and advocacy; population issues
IDENTIFICATION: American biologist and environmental philosopher
BORN: May 29, 1932; Philadelphia, Pennsylvania
SIGNIFICANCE: Ehrlich has published several books that have been influential in raising awareness and promoting action concerning such problems as the dangers of overpopulation and the possible effects of nuclear war.

Paul Ralph Ehrlich, the son of William and Ruth Ehrlich, displayed an early interest in nature study. Following high school, he enrolled at the University of Pennsylvania, from which he graduated in 1953 with a zoology degree. Ehrlich conducted his graduate work at the University of Kansas, earning his M.A. in 1955 and his Ph.D. in 1957. His doctoral research was in the field of entomology, and his first published book dealt with identification of butterflies. He married Anne Howland in 1954; they had one child, a daughter.

After he received his doctorate in 1957, Ehrlich worked as a research associate on various studies. He joined the faculty of Stanford University's Biology Department in 1959 and became a full professor there in 1966. During this time his interest in ecology and conservation developed more fully. He was promoted to the position of Bing Professor of Population Studies at Stanford in 1976.

Ehrlich is best known to the general public for his vigorous support of conservation, including what some consider radical ideas for preserving the earth's resources. Foremost among the topics he has addressed is the potentially devastating effect that increased human consumption could have on the environment. When his book *The Population Bomb* first appeared in 1968, it was widely circulated and caused much discussion regarding worldwide population growth. In the book Ehrlich maintained that increased population, coupled with decreased food production, would, in the following few decades, result in billions of deaths from starvation. This prediction, however, did not come true—although population growth is increasing, food production, with the help of modern agricultural technology, has increased at an even faster rate. Ehrlich suggested in *The Population Bomb* that the government should place limitations on the number of children a couple could have and went so far as to suggest forced vasectomies for men in overpopulated countries with high birthrates.

Ehrlich's suggestions were not limited to population control. He criticized developed countries, especially the United States, for unwise and excessive consumption of natural resources. He predicted that as resources became depleted, inflation would follow. Developing countries, unable to afford even basic necessities, would suffer the most. He urged the U.S. government to pass legislation mandating limited consumption of natural fuels, proper treatment and disposal of wastes, and extensive conservation of fish and wildlife areas.

In his writings, Ehrlich has also predicted increases in air pollution, ozone depletion, species extinctions, and numbers of deaths caused by acquired immunodeficiency syndrome (AIDS); decreases in food production because of poor farming practices; and growing disparity between rich and poor unless corrective measures were taken. His predictions have increased public awareness of environmental problems, and this increased awareness has had some influence on efforts by government officials and grassroots organizations to address these concerns.

In the years after *The Population Bomb* appeared, Ehrlich wrote several other books—many coauthored with his wife—detailing his concerns for conservation and ecological restraint. Among these are *The Population Explosion* (1990), *Healing the Planet* (1991), *Betrayal of Science and Reason* (1996), and *The Dominant Animal: Human Evolution and the Environment* (2008).

Gordon A. Parker

FURTHER READING

De Steiguer, J. E. "Paul Ehrlich and *The Population Bomb*." In *The Origins of Modern Environmental Thought*. Tucson: University of Arizona Press, 2006.

Ehrlich, Paul R., and Anne H. Ehrlich. *One with Nineveh: Politics, Consumption, and the Human Future*. Washington, D.C.: Island Press, 2004.

Simmons, Ian G. "Paul Ehrlich, 1932-   ." In *Fifty Key Thinkers on the Environment*, edited by Joy A. Palmer. New York: Routledge, 2001.

Endangered Species Act

CATEGORIES: Treaties, laws, and court cases; animals and endangered species

THE LAW: U.S. federal law designed to protect species threatened with extinction

DATE: Enacted on December 28, 1973

SIGNIFICANCE: The Endangered Species Act is an important part of the movement toward the protection of biodiversity in the United States. The legislation has been amended frequently since 1973, but its fundamental purpose—the preservation of species—has remained the same.

The U.S. Congress first demonstrated concern for the conservation of species in the Lacey Act of 1900, which prohibited the transportation in interstate commerce of any fish or wildlife taken in violation of national, state, or foreign laws. Following the extinction of the passenger pigeon, the Migratory Bird Treaty Act of 1918 authorized the secretary of the interior to adopt regulations for the protection of migratory birds.

In the Endangered Species Preservation Act of

1966, Congress declared that the preservation of species was a national policy. The statute authorized the interior secretary to identify native fish and wildlife threatened with extinction and to purchase land for the protection and restoration of such species. The Endangered Species Conservation Act of 1969 further empowered the secretary to list species threatened with "worldwide extinction" and prohibited the importation of any listed species into the United States. The only species eligible for the list were those threatened with complete extinction. Although the 1966 and 1969 statutes did not include any penalties for destroying species on the list, the legislation was the most comprehensive of its kind to have been enacted by any nation.

In legislative hearings in 1973, it was reported that species were being lost at the rate of about one per year and that the pace of disappearance seemed to be accelerating, with potential damage to the total ecosystem. The majority of the members of Congress concluded that it was necessary to stop a further decline in biodiversity, and they passed the Endangered Species Act (ESA), which President Richard Nixon signed into law on December 28, 1973.

The ESA outlines a process for the listing of protected species, authorizes appropriate regulations, and provides for state subsidies and funding for habitat acquisition. The act provides that any species of wild animals or plants may receive federal protection whenever the species has been listed as "endangered" or "threatened." The statute defines "endangered" to mean that the species is currently in danger of becoming extinct within a significant geographical region. The term "threatened" means that the species probably will become endangered within the near future. The definition of a "species" includes any subspecies or any distinct population that interbreeds within a specific region. Species found only in other parts of the world are eligible for inclusion on the U.S. list. The only creatures not eligible for inclusion are those insects that are determined to pose an extreme risk to human welfare.

The ESA makes it a federal offense to take, buy, sell, or transport any portion of a threatened or endangered species. Listed animals, however, may be taken in defense of human life, and Alaska Natives are allowed to use listed animals for subsistence purposes. Additional exemptions may be granted for special cases involving economic hardship, scientific research, or projects aimed at the propagation of species. Individuals may be fined ten thousand dollars for each violation of the law committed knowingly and one thousand for a violation committed unknowingly. Harsher criminal penalties are available in extreme cases.

The ESA assigned most enforcement and regulatory powers to the heads of two executive departments. The secretary of commerce, through the National Marine Fisheries Service (NMFS), has responsibility for threatened and endangered marine species. The secretary of the interior exercises formal responsibility for the protection of other species, but the secretary delegates most of the work to the U.S. Fish and Wildlife Service (FWS), which is assisted by the Office of Endangered Species (OES).

In order to benefit from the ESA, species must be officially designated as either endangered or threatened. The courts have consistently ruled that the ESA cannot be used to protect an unlisted species. Species may be proposed for listing by the NMFS, the FWS, private organizations, or citizens. Species are listed only after

---

Nixon on the Endangered Species Act

President Richard M. Nixon made the following statement upon signing the Endangered Species Act of 1973 into law:

At a time when Americans are more concerned than ever with conserving our natural resources, this legislation provides the Federal Government with needed authority to protect an irreplaceable part of our national heritage—threatened wildlife.

This important measure grants the Government both the authority to make early identification of endangered species and the means to act quickly and thoroughly to save them from extinction. It also puts into effect the Convention on International Trade in Endangered Species of Wild Fauna and Flora signed in Washington on March 3, 1973.

Nothing is more priceless and more worthy of preservation than the rich array of animal life with which our country has been blessed. It is a many-faceted treasure, of value to scholars, scientists, and nature lovers alike, and it forms a vital part of the heritage we all share as Americans. I congratulate the 93d Congress for taking this important step toward protecting a heritage which we hold in trust to countless future generations of our fellow citizens. Their lives will be richer, and America will be more beautiful in the years ahead, thanks to the measure that I have the pleasure of signing into law today.

comprehensive investigations have been conducted; open hearings are also held, and opportunities are offered for public involvement in these decisions.

The first list of endangered species, published in 1967, included 72 species. By 1976 the list had grown to 634 species. As of 1995, 1,526 species of plants and animals were listed, including more than 500 that were foreign, and almost 4,000 candidate species were awaiting a listing determination. By 2010, a total of 1901 plants and animals were listed. Although the FWS is required to prepare a recovery plan for each listed species, only a few have recovered sufficiently to be taken off the list.

The act requires that critical habitat for threatened or endangered species be designated whenever possible. All federal agencies have special obligations to determine whether their projects or actions jeopardize the continued existence of a species. Following the U.S. Supreme Court's controversial ruling in *Tennessee Valley Authority v. Hill*, Congress passed the ESA amendments of 1978, which allow consideration for economic factors in the designation of critical habitat. Especially controversial is the section of the act requiring the FWS to formulate and enforce regulations on private lands that provide habitat for listed species. The government must compensate owners in those rare cases when regulations eliminate almost all productive and economic uses of their property, but not when landowners continue to have partial productive use of their land.

Many people who live in rural regions of the United States, particularly in the West, have been highly critical of the ESA, charging that it causes significant job losses as it protects minor subspecies, such as the northern spotted owl. In 1995-1996, a conservative coalition of Republican members of Congress tried to pass a bill that would have weakened the ESA. The controversy that ensued, however, demonstrated that the existing law enjoyed considerable support, and the proposed bill was never passed. Most experts agree that the economic impact of the ESA on the national economy is minimal, but the act does cause hardship for small landowners in some instances. Many environmentalists would support revisions of the law that would give less emphasis to particular species and place more concern on the need for sufficient habitat to support healthy biodiversity, but others fear that such complexity would make the law ineffective.

Thomas T. Lewis

FURTHER READING

Baur, Donald C., and William Robert Irvin. *The Endangered Species Act: Law, Policy, and Perspectives*. Chicago: ABA Publishing, 2002.

Burgess, Bonnie B. *Fate of the Wild: The Endangered Species Act and the Future of Biodiversity*. Athens: University of Georgia Press, 2001.

Goble, Dale D., J. Michael Scott, and Frank W. Davis, eds. *Renewing the Conservation Promise*. Vol. 1 in *The Endangered Species Act at Thirty*. Washington, D.C.: Island Press, 2005.

Kohm, Kathryn, ed. *Balancing on the Brink of Extinction: The Endangered Species Act and Lessons for the Future*. Washington, D.C.: Island Press, 1991.

Mann, Charles, and Mark Plummer. *Noah's Choice: The Future of Endangered Species*. New York: Knopf, 1995.

Noss, Reed, Michael O'Connell, and Dennis Murphy. *Habitat Conservation Under the Endangered Species Act*. Washington, D.C.: Island Press, 1997.

Regenstein, Lewis. *The Politics of Extinction*. New York: Macmillan, 1975.

Rohlf, Daniel. *The Endangered Species Act: Protection and Implementation*. Stanford, Calif.: Stanford Environmental Law Society, 1989.

Scott, J. Michael, Dale D. Goble, and Frank W. Davis, eds. *Conserving Biodiversity in Human-Dominated Landscapes*. Vol. 2 in *The Endangered Species Act at Thirty*. Washington, D.C.: Island Press, 2006.

Environmental law, U.S.

CATEGORIES: Treaties, laws, and court cases; preservation and wilderness issues; resources and resource management; human health and the environment

DEFINITION: Federal and state legislation regulating uses of the environment

SIGNIFICANCE: Although a relatively recent aspect of the American legal landscape, environmental law has come to play a major role in the use of natural resources. Environmental legislation is often controversial in the political arena, but there is widespread popular support among Americans for protecting the environment through statutes and regulation.

At the origin of the American legal system no body of laws existed that directly regulated the environment. Over the course of the twentieth century, however, the nation built up a body of state and federal laws designed to conserve, protect, and restore the environment. Several regulatory agencies are entrusted with the authority to administer these laws. Although there is a large body of environmental law in twenty-first century United States, it is not easy to navigate, as it is divided among federal, state, and even international jurisdictions, is made up of numerous statutes across various legal codes, and is administered in pieces by a host of administrative agencies. For example, the Occupational Safety and Health Act of 1970, which addresses workplace safety issues, has numerous provisions relating to environmental concerns. This law's provisions are administered by both the Department of Labor and the Environmental Protection Agency (EPA).

Conservation Laws

Sustained environmental legislation in the United States began as efforts to conserve public land. Under early common law, the right to property was considered nearly absolute, even when it included the extravagant consumption of natural resources. The influential eighteenth century British jurist William Blackstone described the rights of a property owner over his or her land as "sole and despotic." The only environmental restraint on use of property was direct damage to the property of another person through the law of nuisance or trespass. Although the emphasis among early American settlers was on bringing natural resources quickly into economic use, there was also a sense, dating from the colonial era, that common lands needed to be conserved. For example, a South Carolina statute from 1671 outlawed the poisoning of waterways with impurities. In 1681 William Penn decreed that for every five trees cut down in his Pennsylvania colony, one had to be conserved. The public trust doctrine obligated states to protect tidal shorelines for the common enjoyment of the public.

With the heavy industrialization of the United States during the late nineteenth century, an awareness began to grow that the nation's natural bounty was becoming endangered. Writers such as Henry David Thoreau and John Muir pointed out the fragile beauty of the land and did much to foster a nascent conservation movement. In 1872 the U.S. Congress established Yellowstone National Park, the first national park, to preserve the land from spoliation. The first federal environmental law is generally considered to be the Rivers and Harbors Act of 1899, which banned pollution of the nation's waterways. The Burton Act of 1905 limited hydroelectric power drawn from Niagara Falls. In 1916 Congress established the National Park Service. Everglades National Park was created in 1947. The Wilderness Act of 1964 allowed for what would eventually be more than 40.5 million hectares (100 million acres) to be set aside as wilderness areas.

By 2010 the United States had created 392 national parks comprising 34 million hectares (84 million acres) of conserved land; 552 national wildlife refuges were also in existence. In addition, various social and economic legislation passed during the Progressive and New Deal eras had beneficial effects on conservation. For example, the Civilian Conservation Corps was created in 1933 as a New Deal public works program with a focus on the conservation of natural resources.

Environmental Protection Legislation

Although the conservation laws noted above were important for the preservation of federal lands, they were not concerned with protecting the environment in general. During the 1970's, however, the environmental movement came of age, so much so that this period is sometimes described as the "environmental decade." Writers such as Marjorie Stoneman Douglas and Rachel Carson had earlier publicized the plight of endangered habitats; the nation was shocked in 1969 when the polluted Cuyahoga River in Cleveland caught fire. The first Earth Day was held in 1970. The American people had increasingly become aware that something had to be done to protect a decaying environment.

Congress responded by passing the National Environmental Policy Act of 1969 (actually signed into law on January 1, 1970), which established a framework for comprehensive supervision of the environment. Under this framework Congress passed wide-ranging legislation that constitutes the core of U.S. environmental law. In 1970 Congress enacted amendments to the 1963 Clean Air Act that established regulations on emissions from factories and automobiles. The amendments set minimum standards of air quality and required industries to meet those standards by reducing conventional industrial pollutants. (In 1990 the Clean Air Act was again significantly amended to include requirements that would cut emission of chlorofluorocarbons, or CFCs, and also reduce acid deposition and acid rain by limiting sulfur dioxide

Enacted U.S. Legislation Relating Directly to Climate Change

Year	Act	Effect
1978	National Climate Program Act	Establishes a nationally coordinated program of climate monitoring and prediction
1990	Global Change Research Act	Funds research into global climate change
1997	Byrd-Hagel Resolution	Expresses the sense of the Senate that the United States should join only those climate change treaties that do not harm the domestic economy and that require developing nations, as well as developed nations, to take action against global warming
2005	Energy Policy Act of 2005	Supports voluntary reductions in carbon-intensive activities and the export to developing nations of technologies to reduce carbon intensity

transmissions.) The Clean Water Act (also known as the Federal Water Pollution Control Act Amendments) of 1972 regulates discharge of pollutants and toxic substances into waterways, as well as the filling of wetland areas.

The 1947 Federal Insecticide, Fungicide, and Rodenticide Act was amended in 1972 to regulate the use of pesticides. The Noise Control Act, also signed into law in 1972, addresses excessive noise. Another environment-related law passed in 1972 is the Marine Mammal Protection Act, which protects endangered sea life such as whales, dolphins, sea otters, and seals. Of wider scope was the Endangered Species Act of 1973, which contains provisions aimed at the maintenance of biological diversity and the protection of animal groups in danger of extinction. The Safe Drinking Water Act of 1974 focuses on waters that are the source of drinking water in the United States, mandating basic standards of safety and quality. The Solid Waste Disposal Act and the Resource Conservation and Recovery Act, both enacted in 1976, regulate the treatment and disposal of hazardous wastes. The Toxic Substances Control Act, also passed in 1976, regulates the commercial use and removal of toxic substances such as asbestos, radon, and polychlorinated biphenyls (PCBs).

As important as the laws that were enacted were the administrative agencies entrusted with the laws' implementation and enforcement. Administrative agencies are a modern phenomenon of the American legal system, created in the twentieth century as governmental entities with a mixture of executive, legislative, and judicial functions. Administrative agencies are delegated the authority to enact regulations implementing congressional legislation, to enforce those regulations, and to adjudicate disputed issues. The agency that has the most direct oversight of environmental law is the Environmental Protection Agency, which Congress established in 1970 to consolidate enforcement of various federal laws and duties relating to the environment.

The EPA has functions relating to air, water, and noise pollution and to the handling of toxic and waste substances. It both sets standards and regulations implementing congressional legislation in these areas and enforces these standards with permits, sanctions, lawsuits, and other remedies. The Department of the Interior oversees the national parks and other vital resources. It plays a crucial role in conserving federal lands, forests, and parks; managing irrigation and supplying fresh water; and protecting marine and land wildlife. The Department of Energy runs programs to promote the use of solar energy and the conservation of fossil-fuel resources and to ensure the proper disposal of by-products of energy development, such as radioactive waste. The Department of Agriculture addresses soil issues in forests and on farms. The Council on Environmental Quality, a division of the executive office of the president, is responsible for coordinating governmental responses to environmental issues.

Restoration, Cleanup, and Litigation

The environmental laws enacted in the 1970's focus on protecting the environment from future harm. Once these laws were in place, legislators turned to the question of repairing and restoring damage that had already been done. The Comprehensive Environmental Response, Compensation, and Liability Act of 1980 (CERCLA) created a mechanism (a Superfund) to fund the cleanup of certain environmental disasters—toxic sites that had been caused by waste products. By 1992 more than twelve hundred toxic sites were being cleaned up under Superfund supervision. In addition, CERCLA allows for criminal liability in certain circumstances, in addition to civil penalties.

In 1989 the *Exxon Valdez* oil tanker spilled 11 million gallons of oil onto Prince William Sound off the coast of Alaska. Litigation in the case of *Exxon v. Baker* over the extent of Exxon's liability, including punitive damages, went on for two decades. In response to the spill, Congress enacted the Oil Pollution Act of 1990.

Most U.S. environmental statutes have specific clauses allowing private citizens and entities to sue for environmental harms. Many of the legal debates surrounding these clauses have involved the concept of standing—that is, who has the right to sue. Standing is a traditional legal concept that indicates which persons have suffered sufficient direct harm such that they can become plaintiffs. A major case involving standing was *Massachusetts v. Environmental Protection Agency* (2007), in which the U.S. Supreme Court permitted a coalition of environmental groups and state attorneys general to bring a lawsuit compelling the EPA to determine whether carbon dioxide emissions are an air pollutant under the Clean Air Act.

Perhaps the paradigm environment-related litigation is that concerning asbestos. Hundreds of thousand of Americans are estimated to have contracted asbestos-related cancers. Although there is no question that the nation owes relief to these victims, the long-running asbestos litigation has been criticized. For example, the Manhattan Institute's Center for Legal Policy asserted in a 2003 report that massive asbestos tort litigation was more suited to enriching lawyers than to compensating victims. The Manhattan Institute estimated that of the $70 billion paid out by companies for asbestos claims, $40 billion had gone to lawyers. Companies were driven into bankruptcy by asbestos lawsuits, some of which had little connection to the original asbestos exposure.

Some commentators continue to express fears that widespread environmental lawsuits and mass tort litigation are a boon to lawyers but costly for everybody else. Another lively debate concerns whether governmental supervision of polluters should consist of mandated design and performance standards or market-based incentives.

STATE AND INTERNATIONAL LAWS

The individual U.S. states have environmental laws that run parallel to federal mandates, and the interaction between federal and state laws is complex. Federal environmental laws have priority, but under most federal environmental legislation, state authorities are delegated the power to supervise environmental standards. In other words, even though federal legislation may set the guidelines, state agencies run the programs. For example, the EPA sets recommended standards for water quality under the Clean Water Act, but the states enact specific standards for their localities. In the absence of federal legislation addressing greenhouse gas emissions, numerous states, California in particular, have passed legislation to restrict such emissions. One of the most significant pieces of interstate environmental legislation is the Great Lakes Compact, a legal agreement among eight states that details the use and management of the Great Lakes water supply.

Environmental law in the United States can also be affected by international agreements, especially as environmental problems are of an increasingly global dimension. The best-known examples are agreements regarding greenhouse gases. The 1992 United Nations Framework Convention on Climate Change, a major international effort to reduce greenhouse gas emissions, was followed by a comprehensive extension, the Kyoto Protocol, which took full effect in 2005. The Kyoto Protocol made binding on signatory nations many previously agreed-upon reductions in greenhouse gas emissions; the United States, however, is not a signatory to the protocol. Three international agreements have influenced U.S. legislation regarding depletion of the earth's ozone layer: the Vienna Convention for the Protection of the Ozone Layer (1985), the Montreal Protocol on Substances That Deplete the Ozone Layer (1987), and the London Amendment to the Montreal Protocol (1990).

Howard Bromberg

FURTHER READING

Black, Brian, and Donna Lybecker, eds. *Great Debates in American Environmental History.* 2 vols. Westport, Conn.: Greenwood Press, 2008

Brooks, Karl. *Before Earth Day: The Origins of American Environmental Law, 1945-1970.* Lawrence: University Press of Kansas, 2009.

Buck, Susan. *Understanding Environmental Administra-*

tion and Law. 3d ed. Washington, D.C. Island Press, 2006.

Lazarus, Richard. *The Making of Environmental Law.* Chicago: University of Chicago Press, 2004.

Milazzo, Paul. *Unlikely Environmentalists: Congress and Clean Water, 1945-1972.* Lawrence: University Press of Kansas, 2006.

Nagle, John. *Law's Environment: How the Law Shapes the Places We Live.* New Haven, Conn.: Yale University Press, 2010.

Salzman, James, and Barton Thompson. *Environmental Law and Policy.* 2d ed. New York: Foundation Press, 2007.

Sowards, Adam. *The Environmental Justice: William O. Douglas and American Conservation.* Corvallis: Oregon State Press, 2009.

Zuckerman, Tod I., and Nancy K. Kubasek. *The History of U.S. Environmental Law.* Durham, N.C.: Carolina Academic Press, 2010.

Environmentalism

CATEGORIES: Philosophy and ethics; activism and advocacy
DEFINITION: Movement devoted to the protection of natural resources from harmful influences
SIGNIFICANCE: The modern movement known as environmentalism has had many successes in a wide variety of areas of concern, including the protection and preservation of natural areas, the conservation of natural resources, the development of alternative sources of fuels, the promotion of the importance of biodiversity, and the reduction of pollution.

Environmentalism entails advocating for and taking part in activities aimed at preserving and protecting the natural environment. Environmentalists support many different ways of achieving their goals. Among the many concerns of environmentalists are the reduction of the pollution of air, soil, and water; the prevention of the introduction of species of plants and animals into ecosystems to which they are not native; and the prevention of the encroachment of human activities into natural areas.

AMERICAN ENVIRONMENTALISM

When early European settlers arrived in the Western Hemisphere, they exploited every resource they found, including the native populations. Such exploitation continued in the American colonies during English control. After the Revolutionary War, most Americans were committed to environmental exploitation and westward expansion. They swiftly harvested forests and quickly exhausted arable land with nutrient-needy crops. The slaughter of wild animals for food and pelts and of whales for oil, ambergris, and other products was rampant. Discoveries of gold, silver, and other precious minerals were rapidly exploited.

In the nineteenth century some Americans became concerned about this trend, and New England Transcendentalists such as Henry David Thoreau and Ralph Waldo Emerson began to write about the value of the natural environment. Gradually public opinion began to shift, and increasing numbers of Americans began to see the indiscriminate exploitation of natural resources as less than admirable. Grassroots organizations began to form in response to local environmental concerns during the late nineteenth century. John Muir, one of the founders of the Sierra Club and an advocate of preservation of forests in the American West, was instrumental in influencing popular opinion for environmental preservation.

Another important figure during this period was Gifford Pinchot. Pinchot, son of a wealthy land speculator and lumberman, learned about sustainable forestry in Europe. He did not favor wilderness preservation for the sake of scenery or landscape; rather, he was concerned with the conservation of forest resources. Pinchot favored federal ownership and management of public lands. He advocated prudent exploitation of existing forest resources, including the replacement of cut trees with new seedlings. Utilitarian use of the environment became popular with President Theodore Roosevelt, who appointed Pinchot as the first chief of the U.S. Forest Service.

Individual Audubon Society organizations formed in various states late in the nineteenth century, when commercial bird hunting was extensive (many birds were killed so that their feathers could be harvested for the fashion industry). Audubon advocacy was directed toward the preservation of game and wild birds and against the importation of nonnative birds. The various Audubon organizations advocated for passage of the Lacey Act, which would prohibit interstate commerce in illegally captured or protected birds. The act was signed into law on May 25, 1900, by President William McKinley. Many state Audubon groups united to form a national organization in 1905.

In 1933, as part of his New Deal programs during the Great Depression, President Franklin D. Roosevelt created the Civilian Conservation Corps (CCC). This work relief program focused on environmental projects aimed at conserving and developing natural resources. Among their other accomplishments, CCC workers planted millions of trees in the Great Plains and the Midwest.

After the end of World War II in 1945, logging increased dramatically in the United States to meet demands for new construction. The widespread availability of cheap energy also promoted tremendous expansion in industry and mechanized agriculture during this period, and little public concern was expressed about pollution. In 1962, however, fears about the effects of toxic chemicals on human beings and the environment quickly followed publication of Rachel Carson's book *Silent Spring*, which discussed the concentration of toxic and radioactive substances in the food chain, especially the effects of the insecticide dichloro-diphenyl-trichloroethane (DDT) on wildlife and on domesticated animals. The demands for change that followed the publication of Carson's book were met with intense opposition by the U.S. chemical industry, especially by large corporations that manufactured pesticides and herbicides. Corporate publicity unsuccessfully attempted to brand Carson as a fanatic. The publication of *Silent Spring* was an important factor leading to the formation of the Environmental Defense Fund in 1967 and ultimately to the creation of the federal Environmental Protection Agency (EPA) in 1970.

The first Earth Day, on April 22, 1970, was a seminal event. Some twenty million Americans participated in this "teach-in" suggested by Democratic U.S. senator Gaylord Nelson of Wisconsin. Increasing numbers of Americans began to see the pollution of air, water, and soil as a threat to human health, and politicians and industry leaders responded to their concerns.

The mantra "reduce, reuse, and recycle" became popular in the English-speaking world during the late twentieth century as environmentalists sought to encourage the conservation of resources and reduction of the amounts of materials that were being deposited in increasingly scarce landfill space. As the twenty-first century began, many environmentalists emphasized the excessive use of energy and consumption of goods in developed countries. They promoted the reduction of energy consumption, which would require individual lifestyle changes such as reductions in air travel and the use of smaller, more fuel-efficient cars. These suggestions were unpopular with many Americans.

ENVIRONMENTAL ADVOCACY

Prior to the 1960's, various American grassroots groups advocated on behalf of wildlife preservation and opposed such activities as lumbering, dam building, and mining in wilderness areas. After the publication of *Silent Spring*, such efforts expanded to include national campaigns aimed at reducing the pollution caused by toxic chemicals from agricultural and industrial sources. As increasing numbers of environmental groups formed on the national level, their repeated use of legal advocacy led to the establishment of the field of environmental law.

Greenpeace, an international environmental watchdog organization, was formed in 1971 in British Columbia and rapidly became known for its confrontational tactics, pitting environmental activists against corporate and government entities. Originally Greenpeace focused primarily on protests against nuclear testing, whaling, and seal hunting, but over time its work evolved to address many other environmental issues as well. In the early years of the twenty-first century, Greenpeace stated that global warming presents the greatest environmental threat to the planet.

By the late twentieth century, Green political parties had become increasingly important carriers of the environmental message. In the United States, the Green Party gained attention with its nomination of Ralph Nader as candidate for U.S. president in 1996; Nader was also the party's candidate in 2000. Twenty-five state Green parties formed the Association of State Green Parties (ASGP) in 1996; the ASGP was replaced by the Green Party of the United States (GPUS), a federation of forty-six state Green parties, in 2001. The GPUS promotes "ten key values" based on "ecological wisdom, social justice, cooperation, and nonviolence." Among the issues the party considers important are "Frankenfoods" (that is, foods created using genetic engineering), corporate farming, inequities between rich and poor, and the problems caused by global warming.

In European countries, increasing numbers of elected local and national offices have been held by Greens, and several different Green parties have been represented in the European Parliament. The European Green parties generally stand for ecological wisdom, social justice, grassroots participatory democracy, and nonviolence, principles similar but not identical to those of the GPUS.

Many environmentalists oppose the use of genetic engineering in plant and animal breeding and have raised objections to the introduction of genetically modified organisms in the production of food products. One controversial example is the use of recombinant bovine somatotropin (rBST), an artificial hormone produced by genetically modified bacteria, to boost milk production in cows. The U.S. Food and Drug Administration (FDA) has stated that the milk of cows treated with rBST is safe for human consumption, but some critics assert that because such milk has a slightly elevated level of insulin-like growth factor (IGF-1), a hormone, it may have detrimental health effects in humans. The European Union, Japan, Australia, New Zealand, and Canada have all deemed rBST to have unacceptable detrimental health effects on cows and have banned its use for humane reasons.

Many environmentalists contend that the claims made for some genetically modified foods are inflated. An example is "golden rice," a genetically modified rice that biosynthesizes beta-carotene (a source of vitamin A) in the grain. This product was engineered to provide vitamin supplementation to the diets of consumers in Africa and Asia, but critics assert that few studies were done on how much vitamin A remains after the rice is cooked. They further argue that a better approach would be to make wider food choices available to people in need, rather than encouraging them to continue their dependence on a rice diet.

Radical environmentalists sometimes take part in criminal activities that are aimed at preventing what they view as harms to the environment; the term often applied to such activities is "ecoterrorism." Some of the tactics used by ecoterrorists, including tree spiking (in which metal spikes are driven into trees to prevent logging with chain saws), are federal crimes in the United States. Among the groups that have been described as ecoterrorist organizations are Earth First!, the Environmental Liberation Front, and the Animal Liberation Front.

Antienvironmentalism

Environmentalism has had a number of well-known critics. Bjørn Lomborg, a Danish academic, was a Greenpeace advocate before he conducted a series of statistical studies of environmentalist claims. His published conclusions were that many claims of impending environmental disaster are grossly overstated.

Within the United States, some individuals and organizations view environmentalists as people opposed to continued technological progress. Author and filmmaker Michael Crichton alleged that environmentalism is a religion. He criticized environmentalists for not using "complexity theory" in environmental management. Crichton was also critical of the idea of global warming, asserting that environmentalists use statistical "tricks" to hide data that contradict the concept that greenhouse gas emissions affect the environment.

Various governments around the world have taken violent actions against environmental activists. An infamous example is the sinking of the Greenpeace ship *Rainbow Warrior* in 1985 in a New Zealand port by operatives of the French General Directorate for External Security. The *Rainbow Warrior* had been shadowing French nuclear vessels to protest nuclear testing in French Polynesia. The French government paid compensation for sinking the ship.

Global Environmentalism and Sustainable Development

Two similar concepts emerged during the 1990's: green development and sustainable development. Green development puts environmental concerns above social and economic concerns. Those who advocate sustainable development call for meeting immediate social, economic, and environmental needs in a way that can be maintained for future generations. Some observers contend that the concept of environmentalism underwent a paradigm shift before the twenty-first century, during which it was replaced by the concept of sustainability.

The promotion of sustainable development is sometimes characterized as an attempt by developed countries to exert "protectionism/paternalism" on less developed regions of the world. Some critics of sustainable development believe that it requires limits to population growth. Other critics argue that development of any kind conflicts with environmentalism.

Instead of funding large (and costly) infrastructure programs, such as building dams, in developing countries, governments and organizations that promote sustainability tend to focus on so-called appropriate technology to provide cheap solutions to everyday problems. Some of these inexpensive efforts, such as the introduction of solar cookers at refugee camps in Sudan's Darfur region, have been relatively successful.

Anita Baker-Blocker

Further Reading

Andrews, Richard N. L. *Managing the Environment, Managing Ourselves: A History of American Environmental Policy.* 2d ed. New Haven, Conn.: Yale University Press, 2006.

Delcourt, Paul, and Hazel Delcourt. *Living Well in the Age of Global Warming: Ten Strategies for Boomers, Bobos, and Cultural Creatives.* White River Junction, Vt.: Chelsea Green, 2001.

Edwards, Andres R. *The Sustainability Revolution: Portrait of a Paradigm Shift.* Gabriola Island, B.C.: New Society, 2005.

Kline, Benjamin. *First Along the River: A Brief History of the United States Environmental Movement.* 3d ed. Lanham, Md.: Rowman & Littlefield, 2007.

Liddick, Don. *Eco-Terrorism: Radical Environmental and Animal Liberation Movements.* Westport, Conn.: Praeger, 2006.

Lomborg, Bjørn. *The Skeptical Environmentalist: Measuring the Real State of the World.* New York: Cambridge University Press, 2001.

Maher, Neil M. *Nature's New Deal: The Civilian Conservation Corps and the Roots of the American Environmental Movement.* New York: Oxford University Press, 2008.

European Green parties

Categories: Activism and advocacy; philosophy and ethics

Definition: European political parties that seek to influence society toward greater consciousness of environmental issues

Significance: The Green movement has been particularly strong in Europe, and the efforts of European Green parties have had important impacts on governments' approaches to environmental issues as well as on policy making in the related areas of economics, social justice, and foreign relations.

Ecological parties, or Green parties, can be found across the globe. In Europe many of these parties promote sustainable development, environmental justice, improvement of the quality of life for all individuals, foreign policy centered on peaceful means, and the reorientation of the European Union to emphasize social and environmental issues rather than focusing solely on economic issues. Green parties are fairly unique among political parties inasmuch as they do not always align themselves easily at one end or another of the left-right political spectrum; they often target issues that fall on both sides. Without actually upsetting the cleavage structures of established parties, Greens add new dimensions with their focus on the environment and social justice.

Most Green parties can trace their origins to the environmental movement of the 1960's, the protest movements of the 1970's, and the peace movement throughout the 1980's. Concerns regarding environmental degradation and the testing of nuclear weapons led to increased support of these movements, but it was not until the 1970's that true political entities were formed around these issues. The first recognized ecology party in the world was established in Great Britain in 1973. Over time the party gained political power, and by 1989 the United Kingdom Greens were able to secure 15 percent of the votes yet were still unable to attain any seats in Parliament. Throughout the 1980's and 1990's more Green parties emerged, and many gained legislative seats. By 2010 more than thirty-five countries throughout Europe had Green political parties; more than sixty-five such parties were in existence worldwide. The most successful Green parties, in terms of seats in legislatures, have been those in Western Europe.

The German Greens and the European Parliament

Although not the first to form or the first to gain political representation, the German Green Party is widely regarded as the mother of all Green parties. The German Greens (Die Grünen) were the first ecological party to gain large-scale representation in the federal parliament. In 1983 they garnered the support of almost one million voters throughout Germany and were awarded 28 seats out of 497, or 5.6 percent. In 2009 the German Green Party received roughly 10.7 percent of the total vote.

The European Parliament operates as a legislative body for Europe. Members of the parliament tend to align on the basis of ideological interests instead of by national identity. Recognizing the benefits of forming political alliances within the European Parliament,

Greens from various nations began to pair with other individuals and parties. In 1984 European Greens formed a coalition with regionalists who favored devolution (the return of powers to the subnational units of government), creating the Rainbow Group. This coalition dissolved in 1989, and the Greens became the Green Group. These two groups later reunited and became the Group of the Greens—European Free Alliance, with a total representation of fifty-five members in the seventh European Parliament (2009-2014).

Other Green Parties

Operating under a majoritarian electoral system, the Green Party in France (Les Verts) received as much as 10.6 percent of the popular vote in the parliamentary elections of 1989. The Irish Green Party (Comhaontas Glas) has been in existence since 1981. Promoting public transportation, an environmentally friendly economy, clean politics, and an honest tax package, the party has managed to maintain roughly 4 percent of the national vote.

The Scandinavian countries tend to favor proportional representation systems, and that has helped their Green parties succeed in gaining representation. In Finland the Green League (Vihreä liitto) has continually increased its number of seats in the parliament. The Danish Greens ran their first campaign in 1985. Their platform focuses on a variety of issues, including the creation of a society free of violence, grassroots democracy, and a desire to end poverty. When the Swedish Green Party (Miljöpartiet de Gröna) gained seats in the parliament in 1988, it was a landmark event, as the Green Party was the first new party to enter the Swedish parliament in seventy years. The scope of the issues addressed by the Swedish party is much broader than that of many other ecological parties; the Swedish Greens emphasize decentralization, direct democracy, social justice, gender equality, and placing the environment before short-term economic interests.

Kathryn A. Cochran

Further Reading

Bomberg, Elizabeth E. *Green Parties and Politics in the European Union.* New York: Routledge, 1998.

Burchell, Jon. *The Evolution of Green Politics: Development and Change within European Green Parties.* Sterling, Va.: Earthscan, 2002.

Carter, Neil. *The Politics of the Environment: Ideas, Activism, Policy.* 2d ed. New York: Cambridge University Press, 2007.

Cassola, Arnold, and Per Gahrton, eds. *Twenty Years of European Greens, 1984-2004.* Brussels: European Federation of Green Parties, 2003.

Dobson, Andrew. *Green Political Thought.* New York: Routledge, 1995.

Hanley, David L. *Beyond the Nation State: Parties in the Era of European Integration.* New York: Palgrave Macmillan, 2008.

Müller-Rommel, Ferdinand, and Thomas Pogunkte, eds. *Green Parties in National Governments.* Portland, Oreg.: Frank Cass, 2002.

Federal Land Policy and Management Act

Categories: Treaties, laws, and court cases; land and land use

The Law: U.S. federal law governing how the Bureau of Land Management manages public lands

Date: Enacted on October 21, 1976

Significance: The Federal Land Policy and Management Act of 1976 is the guiding law for the development, enhancement, and protection of designated public lands in the United States. The act mandates that the Bureau of Land Management administer these lands in a sustainable way to ensure their use for generations to come.

In 1976, as a result of years of neglect of public lands that had allowed the proliferation of a number of problems, such as vandalism and destruction of natural resources, lack of sanitation facilities, littering, and overuse, the U.S. Congress enacted the Federal Land Policy and Management Act (FLPMA). For the first time a law provided jurisdiction for the management of public lands under one federal government agency, the Bureau of Land Management (BLM), which operates within the Department of the Interior. Prior to the passage of FLPMA in 1976, the BLM was managing public lands under a number of different laws; the new legislation gave the BLM a unified way of managing public lands, which are defined as lands that are owned by the federal government, excluding lands that are controlled by Native American tribes or set aside for national forests, national parks, and military uses.

FLPMA repealed many obsolete laws related to the management of public lands and gave the BLM new tools for administering such lands. One important purpose of the law is to enable the federal government to retain ownership of public lands while allowing some exchanges of lands and even sales in specific cases. The act's policy declarations specify, among other things, that public lands and their resources must be inventoried periodically and systematically; that all lands that have not previously been designated for any specific uses must be reviewed; that the lands shall be managed on a multiple-use basis as guided by public land-use planning; that they shall be managed in such a way as to protect the quality of their scientific, scenic, historical, ecological, environmental, and archaeological values; and that they shall be managed in a manner that recognizes the nation's need for domestic sources of food, fiber, timbers, and minerals.

FLPMA also states that the federal government is to receive fair market value for the use of public lands and their resources, and that it shall provide payments to compensate state and local governments for any burdens created as a result of the immunity of federal lands from state and local taxation. The act includes provisions covering the disposal of public lands, the acquisition of nonfederal lands, and exchanges of public and private lands; it also addresses regulations concerning the protection of public land areas that have critical environmental concerns.

Public lands in the United States total approximately 105.2 million hectares (260 million acres), the majority of these being in the West. These lands represent approximately 40 percent of the federally owned land, 12 percent of the U.S. land area, and 20 percent of the land situated between the Rocky Mountains and the Pacific Ocean. Most of these lands are located in the states of Nevada, Utah, Wyoming, Idaho, and Oregon. Among other uses, public lands support grazing for livestock on more than 55.4 million hectares (137 million acres) in eleven western states.

Lakhdar Boukerrou

FURTHER READING

Allen, Leslie. *Wildlands of the West: The Story of the Bureau of Land Management.* Washington, D.C.: National Geographic Society, 2002.

Loomis, John B. *Integrated Public Lands Management: Principles and Applications to National Forests, Parks, Wildlife Refuges, and BLM Lands.* 2d ed. New York: Columbia University Press, 2002.

Skillen, James. *The Nation's Largest Landlord: The Bureau of Land Management in the American West.* Lawrence: University Press of Kansas, 2009.

Fish and Wildlife Act

CATEGORIES: Treaties, laws, and court cases; resources and resource management; animals and endangered species
THE LAW: U.S. federal legislation regulating fish and wildlife resources
DATE: Enacted on August 8, 1956
SIGNIFICANCE: Although the Fish and Wildlife Act focused on the commercial fishing industry in the United States, it also laid out protections for sportfishing and expanded general public opportunities for access to fish and wildlife resources.

The Fish and Wildlife Act of 1956 established the U.S. Fish and Wildlife Service within the Department of the Interior; the act also created the positions of director and assistant secretary of the Fish and Wildlife Service, with both positions to be appointed by the president of the United States with Senate approval. The act focused on the commercial fishing industry and provided administration to make sure that American citizens would maintain the right to fish for recreational purposes. The act created the Bureau of Commercial Fisheries, which would oversee the $10 million Fisheries Loan Fund. This fund would be used to invest in operations relating to fisheries; to provide for preservation, restoration, and equipment for fishing vessels; and to study concerns in the fisheries themselves.

Annual research money of up to $5 million was another provision of the act. The reports generated by the research would cover topics such as the domestic production of fish and fish goods, the foreign production of such goods as they affect American industries, the biological information needed to understand these industries, and the creation of more fish and wildlife refuges. Also to be funded were investigations into the impacts of pesticides, fungicides, insecticides, and herbicides on fish and wildlife, including determination of the levels of such chemicals that could be dangerous.

The act was implemented to reserve and plan for the proper management of the valuable renewable resources of fish, shellfish, and wildlife, which contribute to the U.S. economy and food supply. Other stated justifications for the act were that it would generate employment opportunities, would fortify the national defense with trained sailors and available ships, and would improve the general health and physical fitness of sportsmen who could serve in times of military necessity. Amendments to the act in the years since its passage have added requirements for the secretary of the interior to develop wildlife refuges and to provide education programs for the public.

Theresa L. Stowell

FURTHER READING

Hathaway, Jessica. "Fifty Years Ago (Fishing Back When)." *National Fisherman*, September, 2006, 8.

McKay, David. "Environmental Politics." In *American Politics and Society*. 7th ed. Malden, Mass.: Wiley-Blackwell, 2009.

Foreman, Dave

CATEGORIES: Activism and advocacy; preservation and wilderness issues

IDENTIFICATION: American environmental activist and author

BORN: October 18, 1946; Albuquerque, New Mexico

SIGNIFICANCE: As one of the cofounders of the radical environmental group Earth First! and through his continued leadership in less radical organizations, Foreman has had a great deal of influence on the environmental movement.

Dave Foreman, the son of Benjamin and Lorane Foreman, was born in 1946. His father was a pilot in the U.S. Air Force, and the family moved often. They were living in Blythe, California, in 1964 when Foreman graduated from high school. He attended junior college for one year and enrolled the following fall at the University of New Mexico in Albuquerque. His involvement in politics began in high school when he did volunteer work for Republican senator Barry Goldwater. In 1966 Foreman was the state chairman of the conservative organization Young Americans for Freedom and a vocal supporter of U.S. involvement in the Vietnam War. In 1967 Foreman enlisted in the Marines Corps, but he found life in the military unsatisfactory, and he fled into the mountains of New Mexico. Eventually he turned himself in and served time in prison; he was dishonorably discharged from the service.

In 1970 Foreman took to the woods again, supporting himself with odd jobs while spending time backpacking and rafting. He joined an environmental organization called the Black Mesa Defense Fund, and his knack for waging effective campaigns against developers was apparent. The Wilderness Society employed Foreman in 1973 as its principal consultant in New Mexico, and he continued to lead campaigns against polluters and land developers. By 1980 he was one of the Wilderness Society's most visible members.

Foreman, however, felt the need to become more confrontational. In 1980 he resigned from the Wilderness Society and cofounded, with several others, Earth First!, a radical environmental group that used direct action to combat environmental destruction. The founders were inspired in part by Edward Abbey's novel *The Monkey Wrench Gang* (1975), which details the exploits of a small group of people who destroy construction equipment to stop environmental destruction in the southwestern United States. One of the early actions conducted by Earth First! was the unfurling of a 300-foot piece of black plastic down the front of Glen Canyon Dam to represent a crack in the structure. In 1985 Foreman published *Ecodefense: A Field Guide to Monkeywrenching*, a volume of detailed instructions on how to destroy heavy machinery, deface billboards, and spike trees (drive long metal spikes into trees to prevent logging with chain saws).

Earth First!, however, developed problems in keeping its aims directed to environmental protection, and in 1989 Foreman left the group. Around the same time he was arrested on charges of conspiring to topple power lines in Arizona. The case was heard in 1991, and Foreman pled guilty to a reduced conspiracy charge with delayed sentencing and an agreement that the charge would be reduced after five years of good behavior. In 1995 Foreman accepted a three-year term as a director of the Sierra Club, but he resigned in 1997 to devote more time to the Wildlands Project, which he had initiated in the early 1990's. The Wildlands Project founded a think tank in 2003, the Rewilding Institute, which has as its stated mission "to develop and promote the ideas and strategies to advance continental-scale conservation in North America."

Kenneth H. Brown

FURTHER READING

Foreman, Dave, and Bill Haywood, eds. *Ecodefense: A Field Guide to Monkeywrenching*. 3d ed. Chico, Calif.: Abbzug Press, 2002.

Scarce, Rik. *Eco-Warriors: Understanding the Radical Environmental Movement*. Updated ed. Walnut Creek, Calif.: Left Coast Press, 2006.

Fossey, Dian

CATEGORIES: Activism and advocacy; animals and endangered species
IDENTIFICATION: American zoologist and author
BORN: January 16, 1932; San Francisco, California
DIED: December 26, 1985; Virunga Mountains, Rwanda
SIGNIFICANCE: Fossey influenced views of animal behavior and the need for animal protection through her writings about the mountain gorillas of Central Africa and her passionate attempts to save the gorillas from poachers.

Dian Fossey was born in 1932, the daughter of George Fossey III, an insurance agent, and Kitty Fossey, a homemaker. When Fossey was six years old her parents divorced, and she grew up with her mother and her stepfather, Richard Price, a building contractor. After high school, Fossey enrolled in a veterinary medicine program at the University of California, Davis, while supporting herself with low-paying jobs. Her academic difficulties with chemistry and physics courses led her to transfer to San Jose State College, where she earned her bachelor's degree in occupational therapy in 1954. After her postcollege clinical training, she became the director of the occupational therapy department at the Kosair Crippled Children's Hospital in Louisville, Kentucky.

Fossey's love for Africa was inspired by a book on mountain gorillas written by American zoologist George Schaller. In 1963 Fossey took a bank loan to finance a seven-week safari trip to Africa. At Olduvai Gorge in Tanzania, she met with anthropologists Mary Leakey and Louis S. B. Leakey, who were involved with a search for hominid fossils. Fossey's first encounter with a mountain gorilla had a tremendous impact on her. After the end of her African trip, she returned to Kentucky and resumed her work with dis-

---

Fossey's First Gorillas

In her memoir Gorillas in the Mist *(1983), Dian Fossey vividly describes a pivotal moment in her life:*

I shall never forget my first encounter with gorillas. Sound preceded sight. Odor preceded sound in the form of an overwhelming musky-barnyard, humanlike scent. The air was suddenly rent by a high-pitched series of screams followed by the rhythmic rondo of sharp *pok-pok* chestbeats from a great silverback male obscured behind what seemed an impenetrable wall of vegetation. Joan and Alan Root, some ten yards ahead on the forest trail, motioned me to remain still. The three of us froze until the echoes of the screams and chestbeats faded. Only then did we slowly creep forward under the cover of dense shrubbery to about fifty feet from the group. Peeking through the vegetation, we could distinguish an equally curious phalanx of black, leather-countenanced, furry-headed primates peering back at us. Their bright eyes darted nervously from under heavy brows as though trying to identify us as familiar friends or possible foes. Immediately I was struck by the physical magnificence of the huge jet-black bodies blended against the green palette wash of the thick forest foliage.

Most of the females had fled with their infants to the rear of the group, leaving the silverback leader and some younger males in the foreground, standing tense with compressed lips. Occasionally the dominant male would rise to chestbeat in an attempt to intimidate us. The sound reverberated throughout the forest and evoked similar displays, though of lesser magnitude, from gorillas clustered around him. Slowly, Alan set up his movie camera and proceeded to film. The openness of his motions and the sound of the camera piqued curiosity from other group members, who then treed to see us more clearly. As if competing for attention, some animals went through a series of actions that included yawning, symbolic-feeding, branch-breaking, or chestbeating. After each display, the gorillas would look at us quizzically as if trying to determine the effect of their show. It was their individuality combined with the shyness of their behavior that remained the most captivating impression of this first encounter with the greatest of the great apes.

abled children. Three years later Leakey arrived in Louisville and convinced Fossey that she should study the gorillas in the wild as part of a long-term expedition. She agreed and, after paying a short visit to Jane Goodall—the British ethologist was studying chimpanzees in Tanzania—during which she learned methods of data collection, Fossey set up her first campsite and work station at Kabara, Zaire (now the Democratic Republic of the Congo).

Fossey managed to approach and study the gorillas in the remote mountain areas for about seven months, until political unrest took over Zaire. On July 10, 1967, she was arrested by armed guards; she was kept in custody for two weeks, during which time she was raped repeatedly. She eventually managed to escape and found refuge in Uganda. In 1970 she enrolled at Cambridge University in England, where she earned her doctorate in zoology six years later. Immediately thereafter Fossey traveled to Rwanda, where she stayed until 1980, when she accepted a visiting associate professorship at Cornell University. She continued to act as the project coordinator at the Karisoke Research Center, which she had founded in Rwanda in September, 1967.

While at Cornell, Fossey became aware of increased poaching of gorillas in Rwanda and the rapid deterioration of the research center. She returned to Karisoke in June, 1983, with the aim of improving the situation. She actively took over management of the center but failed to keep the support of the National Geographic Society. Her incessant, passionate antipoaching activities created many enemies. On December 26, 1985, Fossey was found dead from machete wounds in her camp in the Virunga Mountains. She was buried in the gorilla cemetery she had built near the camp. Wayne Richard McGuire, an American wildlife researcher, was prosecuted as the primary suspect in her death but fled Rwanda.

Fossey's life was driven by the fighting of poachers and continuation of mountain gorilla conservation efforts that covered a period of almost twenty years in the mountains of Zaire, Uganda, and Rwanda. Poachers, slaughters, and revolutions, as well as loneliness, were part of her daily routine. At the Karisoke Research Center she studied more than fifty gorillas that she described as rather peaceful, charging at humans only when threatened. Fossey was acknowledged as the world's leading authority on the mountain gorillas after the 1983 publication of her book *Gorillas in the Mist*, which dramatically enlarged the contemporary knowledge of gorilla habits, communication, and social structure. Fossey believed that gorillas are altruistic and regal animals whose family structure is unbelievably strong.

Soraya Ghayourmanesh

Further Reading

De la Bédoyère, Camilla. *No One Loved Gorillas More: Dian Fossey—Letters from the Mist*. Washington, D.C.: National Geographic Society, 2005.

Fossey, Dian. *Gorillas in the Mist*. 1983. Reprint. New York: Mariner Books, 2000.

Franklin Dam opposition

CATEGORIES: Activism and advocacy; preservation and wilderness issues

IDENTIFICATION: Resistance to the building of a hydroelectric dam proposed for the Franklin and Gordon river system in southwestern Tasmania, Australia

DATES: 1978-1983

SIGNIFICANCE: In stopping the construction of a hydroelectric dam that promised an industrial boon but threatened to destroy thousands of square miles of Tasmania's temperate rain forest, a grassroots coalition of environmental activists achieved through the Australian court system the first significant legal victory in international efforts to protect natural reserves against development.

In October, 1978, Tasmania's powerful energy conglomerate, the Hydro-Electric Commission, announced ambitious plans to build four dams along the Franklin and Gordon rivers to boost Tasmania's entire energy output by more than 20 percent, a bold move that supporters claimed would guarantee thousands of jobs and catapult the tiny Australian island state of Tasmania into a major industrial economy. The reaction from local environmentalists, however, was immediate. Just five years earlier, the same coalition, known as the Tasmanian Wilderness Society, had failed to stop the construction of a similar dam that had, in turn, destroyed Lake Pedder. The dam proposed for the Gordon River would flood the Franklin River, a wild and barely charted river, as well as thousands of hectares of ancient unspoiled wilderness.

By mid-1980, the antidam coalition was sufficiently organized to stage a massive rally that paralyzed Hobart, Tasmania's capital. The sitting Labor government, which generally supported the project, attempted a compromise, proposing that the dam be moved downriver. The effect, the environmentalists argued, would be the same: the obliteration of the Franklin River basin. Their cause was made more urgent when, in early 1981, caves were discovered in the basin that dated back eight thousand years and contained artifacts and cave paintings made by ancient Aborigines.

A national referendum in late 1981 did little to quash the controversy—voters were asked only to choose between sites rather than to decide on the dam idea itself. When Tasmanians subsequently elected a pro-dam Liberal government, it appeared the dam would be constructed. Conservationists, however, took the fight to the Australian mainland, where they staged a savvy media campaign, citing the Franklin River wilderness area's recent designation as a World Heritage Site by the United Nations Educational, Scientific, and Cultural Organization (UNESCO); World Heritage Sites are cultural and natural sites deemed invaluable to the global community. With bulldozers moving into position, the conservationists organized a blockade of the planned site of the dam that ran from November, 1982, to June, 1983. The blockade created an international outcry as more than one thousand protesters were arrested (to flood Tasmania's limited prison system, the arrested protesters refused bail).

The turning point came when the charismatic Bob Hawke was elected Australia's prime minister in March, 1983. The new Labor government, citing the World Heritage Site designation as an overriding mandate from the international community, ordered work on the dam to stop. This action raised thorny constitutional questions about the reach of the federal government, and the Tasmanian government challenged the federal government's authority in Australia's High Court.

On July 1, 1983, the High Court handed down a narrow (four-to-three) victory to the commonwealth government (and by extension to the environmentalists), finding that the federal government had the power to act to enforce any international treaty. The decision ended efforts to dam the Franklin and Gordon river system, and the campaign to stop the dam came to be regarded as the first successful campaign conducted by what over the next decade would become the Green movement.

Joseph Dewey

FURTHER READING

Buckman, Gary. *Tasmania's Wilderness Battles: A History.* Sydney: Allen & Unwin, 2008.

Harding, Dennis, and Michelle Dale. *Wilderness in Tasmania: The Untouched Land.* Talbot, Tasmania: Tasmanian Book, 2000.

Lines, William J. *Patriots: Defending Australia's Natural Heritage.* St. Lucia: University of Queensland Press, 2006.

McCully, Patrick. *Silenced Rivers: The Ecology and Politics of Large Dams.* Enlarged ed. London: Zed Books, 2001.

Friends of the Earth International

CATEGORIES: Organizations and agencies; activism and advocacy
IDENTIFICATION: International network of environmental organizations
DATE: Established in 1969
SIGNIFICANCE: Friends of the Earth International, the world's largest grassroots environmental network, focuses on the economic and development aspects of sustainability. With its membership weighted toward groups in the developing world, the organization brings global attention to issues such as economic justice, climate justice, and food sovereignty.

Friends of the Earth was founded in the United States in 1969 by David Brower after his departure from the Sierra Club because of that organization's reluctance to challenge the construction of nuclear power plants. Friends of the Earth became Friends of the Earth International (FOEI), an international network, in 1971 following a meeting of like-minded environmental activists from the United States, the United Kingdom, Sweden, and France. By 2010 it had grown into an international network with affiliates in seventy-seven countries and more than five thousand local activist groups.

Unlike many other global nongovernmental organizations dealing with environmental problems and issues, FOEI is structured from the bottom up as a confederation of national organizations or groups. The national organizations are themselves multitiered networks comprising grassroots activists as well as national pressure groups engaged in campaigns for various environmentally progressive and sustainable policies. Activists and groups at all levels are involved in educational activities and research projects.

National and local groups and organizations affiliated with FOEI are required to act independently of established political parties, religious organizations, and any other associations. They are expected to be open and democratic and to uphold nondiscriminatory policies in their practices and internal structures. They are encouraged to cooperate and work with other like-minded organizations pursuing the same goals. National affiliates of FOEI typically work on issues affecting their own countries; they may also choose to participate in any of the international campaigns of FOEI that they deem relevant or important to them. Similarly, local, grassroots activists may choose to work on local, national, or international issues, as they see fit.

National affiliates of FOEI may choose to name themselves Friends of the Earth followed by country name (in parentheses), such as Friends of the Earth (U.S.) and Friends of the Earth (Canada), or they may alternatively use an equivalent translation in the national language, such as Les Amis de la Terre (France) and Amigos de la Tierra (Spain and Argentina). About half of the member groups work under their own names, reflecting thereby an independent foundation and subsequent joining of the FOEI network, such as WALHI (FOE Indonesia), ERA (FOE Nigeria), and NSCN (FOE Norway). Some groups that use the name Friends of the Earth are not members of FOEI; for example, Friends of the Earth (Hong Kong) is not affiliated with FOEI.

FOEI is supported by a secretariat based in Amsterdam and an executive committee that is elected by all member groups at a general assembly held every two years. All overall policy and priority decisions are made at the general assembly.

FOEI views environmental issues in their social, political, and human rights contexts. Its campaigns and activities are typically situated beyond the traditional area of the conservation movement and seek to address the economic and development aspects of sustainability. Although FOEI originated in the United States and Europe, its membership has become more heavily weighted toward groups in the developing world. Some of the organization's priorities in the early twenty-first century include economic justice, climate justice and alternative energy, food sovereignty, and biodiversity.

All of FOEI's international campaigns incorporate elements of three core themes: protection of human and environmental rights, protection of biodiversity, and repayment of the ecological debt owed by rich, industrialized countries to those they have exploited. FOEI is particularly engaged in work on the issues of desertification, protection of Antarctica, water quality and distribution, maritime mining, and nuclear power.

Nader N. Chokr

FURTHER READING

Carter, Neil. *The Politics of the Environment: Ideas, Activism, Policy.* 2d ed. New York: Cambridge University Press, 2007.

Friends of the Earth. *Climate Change: Voices from Communities Affected by Climate Change.* Washington, D.C.: Author, 2007.

_____. *Climate Debt: Making Historical Responsibility Part of the Solution.* Washington, D.C.: Author, 2005.

Porritt, Jonathon, ed. *Friends of the Earth Handbook.* New ed. London: Macdonald Optima, 1990.

Gibbons, Euell

CATEGORIES: Activism and advocacy; forests and plants; agriculture and food
IDENTIFICATION: American ethnobotanist and nature writer
BORN: September 8, 1911; Clarksville, Texas
DIED: December 29, 1975; Sunbury, Pennsylvania
SIGNIFICANCE: Gibbons improved the public image of wild food foraging and thus of environmentalism in general, as his staid, avuncular image made environmentalism acceptable to Americans who had tended to perceive environmental activism as subversive.

By becoming a best seller, Euell Gibbons's first book, *Stalking the Wild Asparagus* (1962), moved the art of collecting and preparing wild foods out of the realm of the eccentric and into the mainstream of popular culture. Gibbons's writing, as well as his lifetime of research both in the field and at the Pendle

Hill Quaker Study Center in Wallingford, Pennsylvania, offered original arguments for the necessity of maintaining the balance of nature. Whereas many environmentalists blamed a fundamentalist reading of the biblical mandate to "subdue the world" for many of the ecological problems of the twentieth century, Gibbons placed equal blame on a misunderstanding of Darwinism. Much of the notion of conquering nature that Gibbons, like most ecologists, despised stemmed, he thought, from misunderstanding of the concept of natural selection. "Survival of the fittest," Gibbons taught, does not always mean "survival of the strongest" but rather "survival of the most cooperative." A forager who uproots a plant is not a destroyer from a larger ecological point of view, since such an action scatters seeds and invariably creates more plants than it destroys. By depicting the human forager as a vital part of the plant's reproductive cycle, Gibbons offered an ideological framework for understanding natural selection without the insistence on predator and prey.

Gibbons's impact on ecological thought went beyond the influence of his books, however. Both the Boy Scouts of America and the U.S. Navy hired him to teach foraging as a survival skill, though Gibbons disliked that context for his discipline. His advocacy of wild foods was culinary rather than survivalist, and he resented the romanticized notion of going into the woods with no provisions and "living off the land." Besides, as a Quaker and pacifist, he resented the military uses to which his skills might be put by the Navy. Nevertheless, he welcomed the opportunity to spread the gospel of cooperation with nature. Another forum for his ideas was a series of seminars he ran at Ithaca College and Bucknell University in the 1970's. With these seminars he assured a wide distribution of his methods by training college students in the first half of the seminar and then guiding them in training high school students in the second half.

Though not formally educated in botany—his college studies were in anthropology and creative writing—Gibbons read virtually every article and book, scholarly and popular, that dealt with the gathering, preparation, and nutritive value of wild plants. He coined the term "ethnobotany" because he saw his field as recovering the botanical and culinary knowledge of Native American culture.

John R. Holmes

FURTHER READING

Cunningham, Anthony B. *Applied Ethnobotany: People, Wild Plant Use, and Conservation.* Sterling, Va.: Earthscan, 2001.

Gibbons, Euell. *Stalking the Wild Asparagus.* 25th anniversary ed. Putney, Vt.: A. C. Hood, 1987.

Gibbs, Lois

CATEGORIES: Activism and advocacy; human health and the environment
IDENTIFICATION: American environmental activist
BORN: June 25, 1951; Buffalo, New York
SIGNIFICANCE: Gibbs united her community by forming the Love Canal Homeowners Association and leading efforts to compel state and federal officials to relocate residents whose homes were compromised by exposure to toxic waste.

In 1977 Lois Gibbs discovered that the elementary school her son was attending in the Love Canal neighborhood of Niagara Falls, New York, was built on top of a toxic chemical dump containing twenty thousand tons of waste. Gibbs was alerted to the presence of the dump in December, 1977, when her son, Michael, began to experience asthma and seizures just four months after he entered kindergarten. She then went door-to-door and questioned other residents about their health in an attempt to determine the full extent of contamination. Two years later, Gibbs traveled to Washington, D.C., to testify on behalf of the Love Canal Homeowners Association at hearings on hazardous waste disposal held before the House Subcommittee on Oversight and Investigations.

Although homes immediately adjacent to the dump were evacuated, residents living outside the "inner ring" remained until January, 1980, when the federal government's Environmental Protection Agency (EPA) financed a controversial scientific investigation by Biogenics Corporation of Houston, Texas. On May 17, 1980, the EPA released a report that stated residents of the area were believed to have damaged chromosomes as a result of exposure to the dump and consequently were at increased risk of cancer, miscarriages, birth defects, and seizures. Gibbs described the report as "one more frightening, scary thing, and we couldn't take it any more. . . . People were very,

very frightened, almost panicked." Shortly afterward, President Jimmy Carter declared a state of emergency at Love Canal and authorized relocation of the residents at a cost of $35 million to the federal government.

Lois Gibbs's fight to evacuate Love Canal captured the public's imagination and was depicted in *Lois Gibbs and the Love Canal*, a 1982 film made for television starring Marsha Mason. The most significant development following the Love Canal disaster was passage of the Comprehensive Environmental Response, Compensation, and Liability Act of 1980 (widely known as Superfund). This legislation created an industry-funded program for cleaning up toxic chemical dumps across the nation. In 1981 Gibbs formed the Citizens Clearinghouse for Hazardous Waste, which was later renamed the Center for Health, Environment, and Justice. This nonprofit organization began by helping community groups suffering from the effects of toxic dumps similar to Love Canal. It eventually expanded its programs to match the growing needs of grassroots environmental organizations, working with more than eight thousand community-based groups.

Peter Neushul

Further Reading

Blum, Elizabeth D. *Love Canal Revisited: Race, Class, and Gender in Environmental Activism*. Lawrence: University Press of Kansas, 2008.

Gibbs, Lois. "What Happened at Love Canal." 1982. In *So Glorious a Landscape: Nature and the Environment in American History and Culture*, edited by Chris J. Magoc. Wilmington, Del.: Scholarly Resources, 2002.

Gore, Al

CATEGORIES: Activism and advocacy; weather and climate
IDENTIFICATION: American environmental activist and politician who served in Congress and as vice president of the United States
BORN: March 31, 1948; Washington, D.C.
SIGNIFICANCE: Through his activism and particularly his participation in the documentary film *An Inconvenient Truth*, Gore has brought worldwide attention to the problem of global warming.

Before serving as the forty-fifth vice president of the United States (1993-2001) under President Bill Clinton, Al Gore was involved in American politics for more than two decades, serving in the U.S. House of Representatives (1977-1985) and then in the Senate (1985-1993) as a representative of Tennessee. Gore is less likely to be remembered primarily for his political service, however, than for his work as an environmental activist of international reputation and impact. From the time of his earliest days in Congress, Gore has been a leading advocate for confronting the threat of global warming. He addressed the issue in his best-selling book *Earth in the Balance: Ecology and the Human Spirit* (1992), and then, as vice president, he led the Clinton administration's efforts to develop economically profitable ways to protect the environment.

After Gore, a Democrat, ran for the U.S. presidency in 2000 and lost to Republican George W. Bush, he began traveling around the United States and internationally to present an educational slide show on global warming; the slide show became the basis of the documentary film *An Inconvenient Truth*, released in 2006. The film won two Academy Awards in 2007, and Gore's message, a warning of the disasters associated with the mounting threat of climate change, won a global audience for the issue. In 2007 Gore was awarded the Nobel Peace Prize, an honor he shared with the Intergovernmental Panel on Climate Change, for "efforts to build up and disseminate greater knowledge about man-made climate change, and to lay the foundations for the measures that are needed to counteract such change."

Gore has attributed the beginnings of his passion for environmental concerns to the influence of his Harvard professor of climate science, Roger Revelle, whose research on climate change ultimately steered Gore toward an interest in politics and government as a means of having a direct impact on environmental policy in the United States. Gore received his bachelor of arts degree in government from Harvard in 1969, then served in the U.S. Army (1969-1971), after which he returned to his studies, attending Vanderbilt University Divinity School (1971-1972) on a Rockefeller Foundation scholarship and then Vanderbilt University Law School (1974), while simultaneously working for the newspaper *The Tennessean* as an investigative reporter (1971-1976).

Gore holds a number of key positions on governing bodies that serve the cause of environmental protection. He is founder and chair of the Alliance for Climate Protection, cofounder and chair of Generation Investment Management, and a partner in the venture capital firm Kleiner Perkins Caufield & Byers, heading that firm's climate change solutions group. Gore has also held faculty positions at Middle Tennessee State University, Columbia University Graduate School of Journalism, Fisk University, and the University of California, Los Angeles.

Gore's critics have often raised four particular issues: They assert that a conflict of interest is suggested by Gore's discordant roles as a private investor in green technology and as a public advocate of taxpayer-funded green-technology subsidies; they point out that his personal level of energy consumption is high; they question the scientific basis of some of his claims; and they object to his refusal to join in open debate with opponents on the subject of global warming. In addition, some organizations, notably the animal rights group People for the Ethical Treatment of Animals (PETA), have criticized Gore for eating meat, a practice that they argue is environmentally unfriendly.

During the 1990's Gore spoke out in opposition to a number of situations that had negative implications for the environment. He opposed the ties of Ronald Reagan's and George H. W. Bush's presidential administrations to Iraqi leader Saddam Hussein, because of the latter's use of poisonous gas and his burgeoning nuclear program. After Saddam's Al-Anfal Campaign, which included nerve-gas attacks on Kurdish Iraqis, Gore cosponsored the Prevention of Genocide Act (1988), legislation that would have cut U.S. assistance to Iraq, but the bill was ultimately defeated.

Wendy C. Hamblet

FURTHER READING

Gore, Al. *Earth in the Balance: Ecology and the Human Spirit.* 1992. Reprint. New York: Rodale Books, 2006.

_____. *An Inconvenient Truth: The Planetary Emergency of Global Warming and What We Can Do About It.* New York: Rodale Books, 2006.

Green movement and Green parties

CATEGORIES: Activism and advocacy; philosophy and ethics

DEFINITIONS: Movement seeking to influence society toward greater consciousness of environmental issues and the political parties established to bring about such change through existing democratic institutions

SIGNIFICANCE: The Green movement seeks to change certain fundamental values in Western society, particularly those that appear to create threats to humanity and the larger nonhuman environment, such as unrestricted technological progress and economic development. Although varied in their origins and interests, Green parties work for political change aimed at achieving the ideals of the Green movement.

The Green movement in Western nations evolved from the various protest movements of the 1960's. Specifically, Rachel Carson's book *Silent Spring* (1962) awakened people to the health hazards of industrial pollution and encouraged them to consider humanity and the environment as interdependent. In addition, the fear that world superpowers might resort to the use of nuclear weapons during the Cold War prompted antinuclear, propeace demonstrations. The peace movement in Europe produced the first formal Green parties in the 1970's, organized on the models of two precursors: New Zealand's New Values Party, started in 1972, and Great Britain's Ecology Party, started in 1973, both of which sought to formulate new electoral strategies with environmental issues.

The Greens first organized in the United States in 1984 as the Green Committees of Correspondence, which took the German Green Party as their model. These organizations consisted of state parties and otherwise unaligned individual members. From the outset, however, Green parties were state controlled; their national organization was a loose confederation. The first individual Green candidate in the United States appeared on a ballot in 1986, and in 1990 the Alaska Green Party was the first to achieve ballot status, followed by the California party in 1992. The coalition of Green parties for the first time fielded a presidential candidate, Ralph Nader, in the 1996 election and attracted 1 percent of the votes. Despite interstate disputes over tactics during the early 1990's, twenty-five state parties formed the Association of State Green Parties (ASGP) in 1996 in order to prepare for national elections in 2000; in 2001, the ASGP was replaced by the Green Party of the United States (GPUS), a federation of forty-six state Green parties.

Some degree of international affiliation exists among national parties. The GPUS, for instance, is a member of the Federation of the Green Parties of the Americas, based in Mexico City, and is associated with the thirty-seven-member European Federation of Green Parties.

Core Principles

Central to the basic tenets of Green parties in all countries is the belief that the world order must be reshaped, with emphasis given to local governance. At the same time, the transformation must involve a shift in people's interest from the immediate future and self-centered satisfaction to the long-range future and sustainable production of material needs. The often-repeated slogan "Think globally, act locally" reflects the spirit of the movement. To this end, the European Greens published four basic principles in the 1970's, to which Greens in the United States added six more in the 1980's.

The "four pillars" of the early European Greens were ecology, social justice, grassroots democracy, and nonviolence. By "ecology" (or "ecological wisdom," the American term) is meant a redirection of mentality: People should consider themselves part of nature, not controllers of it, and so live in harmony with it. Practically, this entails devoting technology to achieving an energy-efficient economy and minimizing extraction of nonrenewable resources. Social justice encompasses universal equal rights, dignity, and social responsibility based on the values of simplicity and moderation. Greens advocate community-controlled, free education and programs to prevent crime. Grassroots democracy, in the American interpretation, would distribute most state and federal power to local elected officials and mediating institutions, such as neighborhood organizations, church groups, voluntary associations, and ethnic clubs. The goal is to restore civic vitality by involving as many people as possible in decision making and avoiding reliance on lawyers, legislators, and bureaucrats. The pillar of nonviolence involves not only seeking to end patterns of conflict in families and communities but also working to eliminate nuclear weapons. Worldwide, Greens opposed the 1991 Persian Gulf War and cited the environmental disaster of burning oil wells and oil spills in the Gulf as evidence that military action is counterproductive at best and likely to be ruinous.

The six key values added by American Green parties elaborate on issues inherent in the four pillars. First, respect for diversity would honor cultural, racial, sexual, and religious differences, while insisting that citizens bear individual responsibility to all beings for their actions. Second, feminism would replace traditions of dominance by a particular ethnic group, social class, or gender with ethics based on cooperation and respect for the contemplative, intuitive capacities of all people. Community-based economics calls for employee ownership of businesses, workplace democracy, and equal distribution of wealth to ensure basic economic security for all. Similar to grassroots democracy, decentralization would give power to economically defined localities and ecologically defined regions; it entails redesigning institutions so that control over regulations and money is greatest at the community level, rather than the national level, and control over environmental policy is greatest at the regional level. Personal and global responsibility encourages wealthy communities to assist grassroots groups in developing countries—directly rather than through their governments—in order to make them self-sufficient. To produce the means for such aid, the Greens want to decrease the national defense budget, although not to the point of compromising American security. Finally, future focus, or sustainability, requires all economic, scientific, and cultural policies to be formulated with careful attention to their long-range effects and not just to their immediate benefits. For this reason, European and American Greens have denounced such scientific developments as genetic engineering and nuclear power.

The European and American versions of the Green movement contain potential conflicts. For example, the European Greens call for internationalized security to prevent war; in particular they want to give control over military action to the United Nations after it is reformed so that each nation has equal voting power. They also want to replace regional trade treaties, such as the North American Free Trade Agreement (NAFTA), with treaties negotiated and monitored by the United Nations. The American goal of decentralization does not clearly accord with this vision of world order, nor is it clear how community-based economics would admit the European goal of planetwide economic solidarity.

Political Power

In the 1980's and 1990's, doctrinal divisions that were already latent among the Greens produced contention and sometimes disaffection, especially in the European organizations. Divisions have tended to fall between moderates, who are open to compromises with other political parties in order to gain power, and radicals, for whom compromise is unacceptable. The moderates, sometimes referred to as "light green," tend to espouse the anthropocentric view: The Greens should help safeguard the human environment, although preferably not at the expense of other organisms. This faction seeks reform of existing social and economic institutions. The radicals, known as "dark green," are biocentric, holding that all creatures have equal natural rights to life and that humankind should not consider itself a favored species. They wish to end the affluent, technological, expansionist, service-providing orientation of society.

Green parties have achieved modest success, especially in Western Europe, in placing political candidates in office. Greens have constituted 10 percent of some parliaments and have entered into coalitions that have formed ruling governments. In nearly all countries, they have influenced environmental legislation. In the United States, by 1999 Green candidates had been elected to sixty-three local offices in fifteen states, mostly for such nonpartisan agencies as planning groups and school boards. Arcata in Northern California was the first municipality to have a Green majority on its city council, and in 2008 a Green was elected to the Arkansas House of Representatives (although he later changed party affiliation, becoming a Democrat).

The power of Green parties has manifested more in influencing policy and in education about environmental and social issues than in direct legislative or administrative action because opposition to the Greens' principles is formidable in most countries. Critics in the United States accuse the Green movement of elitism, asserting that it is the project of well-educated, middle-class white people. The Green reconception of society appears to leave little room for individualism, which also runs counter to mainstream American sensibilities, and the proposal to give international organizations control over security, particularly the United Nations, raises fears among nationalists in all countries that cultural identity and sovereignty will be lost. Perhaps most of all, the citizens of Western nations generally resist the ideas of the redistribution of wealth and sustainability, which would dismantle free market economies. Nonetheless, the Green movement has succeeded in a primary objective: to end the status of technological, economic expansion as an unquestioned value and place the burden of proof on its proponents to demonstrate that specific projects will not harm humanity or nature.

Some mainstream parties, such as the Democrats in the United States, complain that Green parties are spoilers, enticing away liberal votes for candidates who stand little chance of winning elections. For this reason, liberal parties have adopted some of the Green environmental goals in their own platforms, weakening the attraction of the Green parties' candidates.

Roger Smith

Further Reading

Audley, John J. *Green Politics and Global Trade: NAFTA and the Future of Environmental Politics.* Washington, D.C.: Georgetown University Press, 1997.

Carter, Neil. *The Politics of the Environment: Ideas, Activism, Policy.* 2d ed. New York: Cambridge University Press, 2007.

Dickerson, Mark O., Thomas Flanagan, and Brenda O'Neill. "Environmentalism." In *An Introduction to Government and Politics: A Conceptual Approach.* 8th ed. Toronto: Nelson Education, 2010.

Dobson, Andrew. *Green Political Thought.* New York: Routledge, 1995.

Johnson, Huey D. *Green Plans: Greenprint for Sustainability.* Lincoln: University of Nebraska Press, 1997.

Kline, Benjamin. *First Along the River: A Brief History of the U.S. Environmental Movement.* 3d ed. Lanham, Md.: Rowman & Littlefield, 2007.

Pepper, David. *Modern Environmentalism: An Introduction.* 1996. Reprint. New York: Routledge, 2003.

Switzer, Jacqueline Vaughn. *Green Backlash: The History and Politics of Environmental Opposition in the U.S.* Boulder, Colo.: Lynne Rienner, 1997.

Greenpeace

CATEGORIES: Organizations and agencies; activism and advocacy; animals and endangered species; nuclear power and radiation
IDENTIFICATION: International watchdog organization dedicated to protecting the environment
DATE: Established in 1971
SIGNIFICANCE: Through its persistent, nonviolent confrontational tactics, Greenpeace applies pressure to governments, organizations, and private corporations that often results in positive changes for the environment while greatly raising public awareness of particular environmental issues.

Greenpeace grew out of the small organization known as the Don't Make a Wave Committee, which was formed in 1971. The members of the group, which was based in Vancouver, British Columbia, had joined together because of their mutual opposition to nuclear testing; they aimed to use nonviolent, creative confrontation to raise public awareness of dangers to the global environment. In September, 1971, twelve volunteers from the organization, now known as Greenpeace, sailed a small boat into the U.S. atomic test zone on the island of Amchitka, off the coast of Alaska, in an effort to stop the United States from conducting a test of a nuclear weapon. The activists believed the bomb's blast could destroy the wildlife haven, cause an earthquake, or create a tidal wave.

Since that first voyage, Greenpeace protesters have often taken their boats into dangerous zones, placing themselves at risk to save whales from harpoons or to block the dumping of toxic or radioactive wastes into the ocean. Within twenty years of its founding, Greenpeace had supporters in thirty countries; by 2010, it had offices in forty-one countries and 2.8 million supporters around the world.

The work of Greenpeace is based on a belief in the importance of the role of individuals and of the visibility of high-profile actions. The group's funding comes mainly through small, individual donations. The stated "core values" of Greenpeace include "'bear[ing] witness' to environmental destruction in a peaceful, non-violent manner" and using "non-violent confrontation to raise the level and quality of public debate" on environmental issues.

Greenpeace's ships travel the world to bring the attention of the news media to local and regional environmental problems, to raise public awareness. The ships have been positioned between whaling fleets and the whales they were trying to catch and have been used to stop the killing of seals. In 1985 the Greenpeace ship *Rainbow Warrior* was bombed and sunk in Auckland Harbor, New Zealand, by French intelligence agents because it was shadowing French nuclear vessels to protest nuclear testing in French Polynesia.

The main governing body of Greenpeace is Greenpeace International, which is based in Amsterdam. Greenpeace International develops and coordinates global strategies and policies with input from Greenpeace regional offices. Over the years the organization has undergone changes and disputes within its leadership, and in 1996 its articles of association were changed to allow for more efficient allocation of resources. Because of the complexity of the issues that Greenpeace addresses, the organization has devoted increasing resources to scientific research, using its ships as mobile laboratories.

Colleen M. Driscoll

Raising Awareness by Making Waves

In his memoir Making Waves: The Origins and Future of Greenpeace *(2001), Greenpeace cofounder Jim Bohlen relates a realization he had following his return from the ultimately aborted but highly publicized first voyage of the* Greenpeace:

The next Monday morning, November 1st, I returned to work at the UBC Forest Products Laboratory. I walked in the front door, checked in with the receptionist, and walked down the glassed-in connecting corridor towards my office. On the way I met one of my colleagues. He was all smiles, congratulating me on the success of the voyage, and ended with, "Well Jim, what are you going to do next?" Suddenly overcome with a great weariness, I replied, "Bob, what are *you* going to do next?"

Those few words summed up the essence of the campaign. Our job had been to make the people of North America aware of the real, and potential, danger of testing nuclear weapons. We imagined massive public opinion raging against the . . . underground blast and bringing about, once and for all, an end to the mad escalation of nuclear weapons. The *Greenpeace* and its crew had done its job.

Further Reading

Bohlen, Jim. *Making Waves: The Origins and Future of Greenpeace.* Tonawanda, N.Y.: Black Rose Books, 2001.

Weyler, Rex. *Greenpeace: How a Group of Ecologists, Journalists, and Visionaries Changed the World.* Emmaus, Pa.: Rodale Press, 2004.

Hansen, James E.

CATEGORIES: Activism and advocacy; weather and climate

IDENTIFICATION: American climate change scientist

BORN: March 29, 1941; Charter Oak Township, Iowa

SIGNIFICANCE: As a prominent climate scientist and activist, Hansen has been an important contributor to increased public awareness of global warming.

James E. Hansen was born in a farmhouse and reared in Denison, Iowa, about 97 kilometers (60 miles) northeast of Omaha, Nebraska. Hansen's family moved to Denison when James was four years of age; his father, who had been a tenant farmer, took up work as a bartender, and his mother worked as a waitress. With a scholarship and money saved from his job delivering newspapers, Hansen attended the University of Iowa, where he majored in mathematics and physics; he graduated summa cum laude in 1963.

Hansen went on to earn a master's degree in astronomy and a doctoral degree in physics at the University of Iowa; he studied for his Ph.D. with the chairman of the physics department, James Van Allen, who discovered the earth-girdling radiation belts named for him. Hansen decided to specialize in the study of the atmosphere of Venus at a time when scientists were discovering that the planet's super-hothouse atmosphere (with temperatures above 454 degrees Celsius, or 850 degrees Fahrenheit) was 95 percent carbon dioxide.

His doctorate completed, Hansen went to work at the National Aeronautics and Space Administration's Goddard Institute for Space Studies (GISS), which is affiliated with Columbia University in New York City. During 1976 Hansen was working as principal investigator on the Pioneer Venus Orbiter when a Harvard postdoctoral researcher asked him to help calculate the greenhouse effect of human-generated emissions in the earth's atmosphere, and soon Hansen was captivated by the subject of global warming. He was appointed director of GISS in 1981, and that same year, with several coauthors, he was the first to use the term "global warming" in a scientific context. During 1988, Hansen went public with warnings about global warming in testimony before the U.S. Senate.

Several times over many years, transcripts of Hansen's warnings were edited before they were released to the public by the government. Each time, Hansen publicly objected to the edits, provoking a public furor. Soon Hansen was making headlines in *The Washington Post* and *The New York Times* as he compared censorship of science to the tactics of Nazi Germany and the old Soviet Union.

Every month, Hansen's laboratory takes the earth's temperature, monitoring ten thousand temperature gauges around the planet. Hansen has stated that he anticipates that the earth will not experience another ice age unless the human race becomes extinct—and even then it would take several thousand years to restore equilibrium to natural cycles that have been disrupted by the burning of fossil fuels. He has advanced an argument that the earth will not be safe until the proportion of carbon dioxide in the atmosphere is reduced to 350 parts per million (it was 387 parts per million in 2009).

Over time, Hansen has become increasingly adamant in his opposition to coal mining and to coal-fired electrical power plants, which are significant contributors to the carbon dioxide emissions linked with human-caused global warming. He has advocated civil disobedience as an oppositional strategy, and he was among thirty-one protesters arrested for obstructing officers and impeding traffic during a West Virginia demonstration against mountaintop coal mining in June, 2009.

Bruce E. Johansen

Further Reading

Hansen, James E. *Storms of My Grandchildren.* New York: Bloomsbury, 2009.

_____, et al. "Climate Impact of Increasing Atmospheric Carbon Dioxide." *Science* 213 (1981): 957-966.

Johansen, Bruce E. "The Paul Revere of Global Warming." *The Progressive*, August, 2006, 26-28.

Hardin, Garrett

CATEGORIES: Activism and advocacy; population issues; resources and resource management
IDENTIFICATION: American ecologist
BORN: April 21, 1915; Dallas, Texas
DIED: September 14, 2003; Santa Barbara, California
SIGNIFICANCE: Through his writings—in particular his widely read 1968 article "The Tragedy of the Commons"—Hardin raised awareness of the environmental problems caused by human overpopulation and overexploitation of resources.

Garrett Hardin was an ecologist who argued strongly for the need to control human population growth. He first received public attention for his ideas in 1968 with the publication of his article "The Tragedy of the Commons" in the journal *Science*. The title of the article came from the concept of the English commons. In a typical medieval English community, all inhabitants could graze their animals on pasture that was held in common by all. The nature of the commons made it possible for users to overexploit the pasture, because any extra animals that one person raised were available for that user's benefit alone, while the grazing resources would be lost to all users of the commons. When the total population is low, any misuse of a commons can be compensated for by a move to a new area. When the total population has increased to the point where no new land is available, the resource held in common will be depleted. The results can be seen in acid rain and ozone depletion (in which the air is treated as a commons), overgrazing (in which the land is treated as a commons), and the whaling and commercial fishing industries (in which the resources of the sea are treated as a commons).

Hardin noted that each problem could be treated as though it were separate and distinct from every other, but he believed that this approach would simply hide the underlying cause of these problems. The pivotal commons is the freedom to breed, particularly when the costs of having children are spread over the entire population and are not restricted to the parents. According to Hardin, human overpopulation is the single root cause that lies behind problems such as excessive resource consumption, war, starvation, so-called ethnic cleansing, poverty, noise pollution, foreign aid, and simple traffic jams.

As explained in his book *The Limits of Altruism* (1977), Hardin believed that every nation has the responsibility to reduce its population to the level it can support using its own resources. Special privileges enjoyed by a few people today should be tolerated if they contribute to national self-reliance in the future. If all nations do not become self-reliant, the resulting population crash will be so devastating that civilization may never recover even if small enclaves of humans survive.

In one of his last books, *Living Within Limits: Ecology, Economics, and Population Taboos* (1993), Hardin compared the United States to a lifeboat in a sea of human misery and concluded that the nation would be able to survive the coming era of resource scarcity only by banning immigration and stabilizing its population. Hardin denied the workability of the "global village" and offered no apology for his apparent lack of compassion. In the absence of human moderation, he believed, nature will once more assert dominance in controlling human population size through disease, starvation, and war.

Gary E. Dolph

FURTHER READING

Hardin, Garrett. "The Tragedy of the Commons." *Science* 162 (December 13, 1968): 1243-1248.

Pepper, David, ed., with Frank Webster and George Revill. *Environmentalism: Critical Concepts*. New York: Routledge, 2003.

Sooros, Marvin S. "Garrett Hardin and Tragedies of Global Commons." In *Handbook of Global Environmental Politics*, edited by Peter Dauvergne. Northampton, Mass.: Edward Elgar, 2005.

Inconvenient Truth, An

CATEGORIES: Activism and advocacy; weather and climate
IDENTIFICATION: Documentary film about climate change
DATE: Released in 2006
SIGNIFICANCE: Winner of the Academy Award for best feature-length documentary, *An Inconvenient Truth* helped to increase public awareness of the issue of global warming.

An Inconvenient Truth is an influential documentary film directed by Davis Guggenheim; it centers on the environmental advocacy work of former U.S. vice president Al Gore, particularly his efforts to inform the public about the dangers posed by climate change. The film documents a lecture and slide show on global climate change presented by Gore in various locations around the world in the aftermath of his failed 2000 campaign for the U.S. presidency. The slide show includes a well-constructed combination of graphs and charts that present the leading edge of climate science in terms easily understood by nonexperts. In his presentation, Gore also shows maps projecting the extent of possible climate change and photos documenting receding glaciers, melting polar ice, softening tundra, and drying lakes.

An effective narrative strategy deployed in the film is the interspersion of autobiographical vignettes that contextualize Gore's motivations for taking on his environmental agenda. The vignettes help to personalize the sometimes cold scientific data, so that when Gore concludes with a call for individual, community, and governmental action on climate change, he is able to make a more personal human appeal than would have been possible if the film had documented the lecture and slide show alone. The topics of the vignettes range from childhood remembrances of Gore's family's farm in Tennessee to the near-fatal traffic accident of his youngest son in 1989 and his experience of losing the presidential election in 2000. Critics of Gore and his message have objected to the film's vignettes, especially the one discussing the lost presidential election, as self-aggrandizement; they have also asserted that some of the graphic extrapolations in the slide show and film are exaggerated for propaganda purposes, although the climate science Gore shares in the film has been substantiated by the peer review process.

Michael Mooradian Lupro

International Convention for the Regulation of Whaling

CATEGORIES: Treaties, laws, and court cases; animals and endangered species
THE CONVENTION: International agreement setting limits on the hunting of whales
DATE: Opened for signature on December 2, 1946

SIGNIFICANCE: Intended to balance the conservation of whales with benefits from their exploitation, the International Convention for the Regulation of Whaling failed to prevent the collapse of many of the world's whale populations. Since its adoption of a ban on all commercial whaling in 1986, the convention has been a focal point for controversy.

The International Convention for the Regulation of Whaling (ICRW) was established to consolidate several previous whaling agreements and to encourage rapid development of the whaling industry of countries ravaged by World War II. The ICRW created the International Whaling Commission (IWC), which has power to amend the ICRW's schedule (a document that limits which whales can be taken and under what conditions), although the IWC cannot amend the ICRW itself.

The foremost controversial issues facing the ICRW have been its inability to prevent the collapse of the world's great whale stocks during the 1950's and 1960's and its inability to force compliance by its signatory states—these problems have continued to serve as backdrops to debates over whaling. While Japan, Norway, and some other nations argue that a sustainable harvest of some whale stocks has robust scientific support, others—led by the United Kingdom, the United States, Australia, and New Zealand—contend that historical overexploitation counsels a more precautionary approach and that the killing of whales is unethical. Dissatisfaction with the ICRW and the whaling moratorium implemented by the IWC in 1986 has encouraged the creation of the rival North American Marine Mammal Commission, which Iceland, Norway, Greenland, and the Faroe Islands established in 1992 so that they might regulate their own whaling operations.

The ICRW is also challenged by noncompliance occurring either explicitly outside IWC guidelines or within the guidelines but against the spirit of the regulations. The most explicit noncompliance occurred when the Soviet Union engaged in illegal whaling for banned stocks during the Cold War. More recently meat from whale species that are banned from harvest by the IWC was allegedly found in the markets of Japan and South Korea.

The IWC allows for three types of exemptions to the whaling moratorium, the uses of which are sometimes interpreted to be against the intent of the moratorium, though within IWC guidelines. First, a member state can file a general reservation to the moratorium, an exemption Norway has used to engage in limited commercial whaling. Second, a scientific exemption allows for the taking of whales for research purposes. Only Japan practices scientific whaling, and it has been accused of using this exemption as a cover for commercial operations. Finally, an exemption for aboriginal subsistence whaling allows harvesting by indigenous peoples with strong historical whaling cultures. Japan, Norway, and Iceland have objected to the exemption for aboriginal subsistence whaling, asserting that it is inconsistent with restrictions placed on their own whaling communities.

Because the ICRW lacks effective enforcement mechanisms, compliance must often be negotiated externally. Notably, Japan agreed to comply with the whaling ban after the United States threatened to bar Japan from lucrative American fishing grounds, and Norway has received similar threats regarding its whaling expeditions. Despite these controversies, the ICRW remains the premier global whaling treaty, and the IWC has extended its scope of responsibility to other kinds of threats to whale populations, such as competition with fisheries, chemical and acoustic pollution, ozone depletion, and global warming.

Adam B. Smith

FURTHER READING

Burns, William C. G., and Alexander Gillespie, eds. *The Future of Cetaceans in a Changing World.* Ardsley, N.Y.: Transnational, 2003.

Gillespie, Alexander. *Whaling Diplomacy: Defining Issues in International Environmental Law.* Northampton, Mass.: Edward Elgar, 2005.

Tønnessen, J. N., and A. O. Johnsen. *The History of Modern Whaling.* Berkeley: University of California Press, 1982.

International Institute for Environment and Development

CATEGORIES: Organizations and agencies; activism and advocacy; resources and resource management
IDENTIFICATION: Nonprofit international environmental research and advocacy organization
DATE: Established in 1971
SIGNIFICANCE: The International Institute for Environment and Development works to address global poverty and to promote the sustainable management of the world's natural resources.

The International Institute for Environment and Development (IIED), which is based in London, was launched in 1971 by Barbara Ward, a world-renowned economist and policy adviser. IIED is considered to be one of the first nonprofit organizations to link the environment with development. The stated mission of the IIED is "to build a fairer, more sustainable world, using evidence, action and influence in partnership with others." Since its creation, IIED has played a key role in shaping major environment-related international conferences, including the 1972 United Nations Conference on the Human Environment, the 1992 Earth Summit, and the 2002 World Summit on Sustainable Development.

IIED's research and development work is centered on addressing the local needs of some of the world's most vulnerable people. The organization's aim is to ensure that these people, the disadvantaged, have a say in the environmental policies that affect them. In collaboration with grassroots partner organizations, IIED works to develop programs that are relevant to the needs of disadvantaged populations. The five major program areas on which IIED focuses are as follows: natural resources, including sustainable agriculture, biodiversity, drylands, and forestry; climate change, including mitigation, adaptation, and vulnerability; human settlements and related areas, such as urban poverty, urban environment, and rural-urban interactions; sustainable markets, especially environmental economics, corporate responsibility, market reorganization, and trade; and governance, local and global, in the areas of law and planning.

Over the years, IIED has developed core concepts and methods for addressing the relationship between sustainable development and social issues. Some of the first ecotaxation and green accounting methodologies, now used by many governments and businesses, were developed by economists at IIED. Consultation with IIED has led a number of companies to improve how they address the environmental problems created by their processes and to develop best management practices to protect and conserve natural resources. IIED has also helped develop methodologies that promote public participation in environment-related decisions; these visual techniques are now widely used by the international and local development communities to encourage the airing of stakeholder ideas and views.

IIED has been involved in shaping the global debate on climate change. The organization's climate change programs focus on the least developed countries and on small island developing states, with particular attention to the economics of climate change adaptations. Many small island countries are vulnerable to the sea-level rises projected as a result of climate change. IIED seeks to provide decision makers with the tools and options they need to take appropriate and effective measures in adapting to climate change.

Lakhdar Boukerrou

International Whaling Commission

CATEGORIES: Organizations and agencies; animals and endangered species
IDENTIFICATION: International body established to regulate whaling
DATE: Established on December 2, 1946
SIGNIFICANCE: Although the International Whaling Commission was created to ensure the conservation of whales through the regulation of whaling, it eventually evolved into a forum serving nations primarily interested in weakening restrictions on whaling.

During the 1930's, concerns regarding declining whale stocks and the near extinction of some species prompted efforts to regulate the whaling industries of the world's nations. These efforts at international cooperation met with little success, but as World War II drew to a close, proponents of regulation pressed for a regulatory framework. In November, 1945, representatives of several whaling nations gathered in Washington, D.C., for the International Whaling Conference. Although some delegates argued for an agency under the direction of the United Nations, the conference led to the creation of an autonomous organization called the International Whaling Commission (IWC) in 1946. Great Britain,

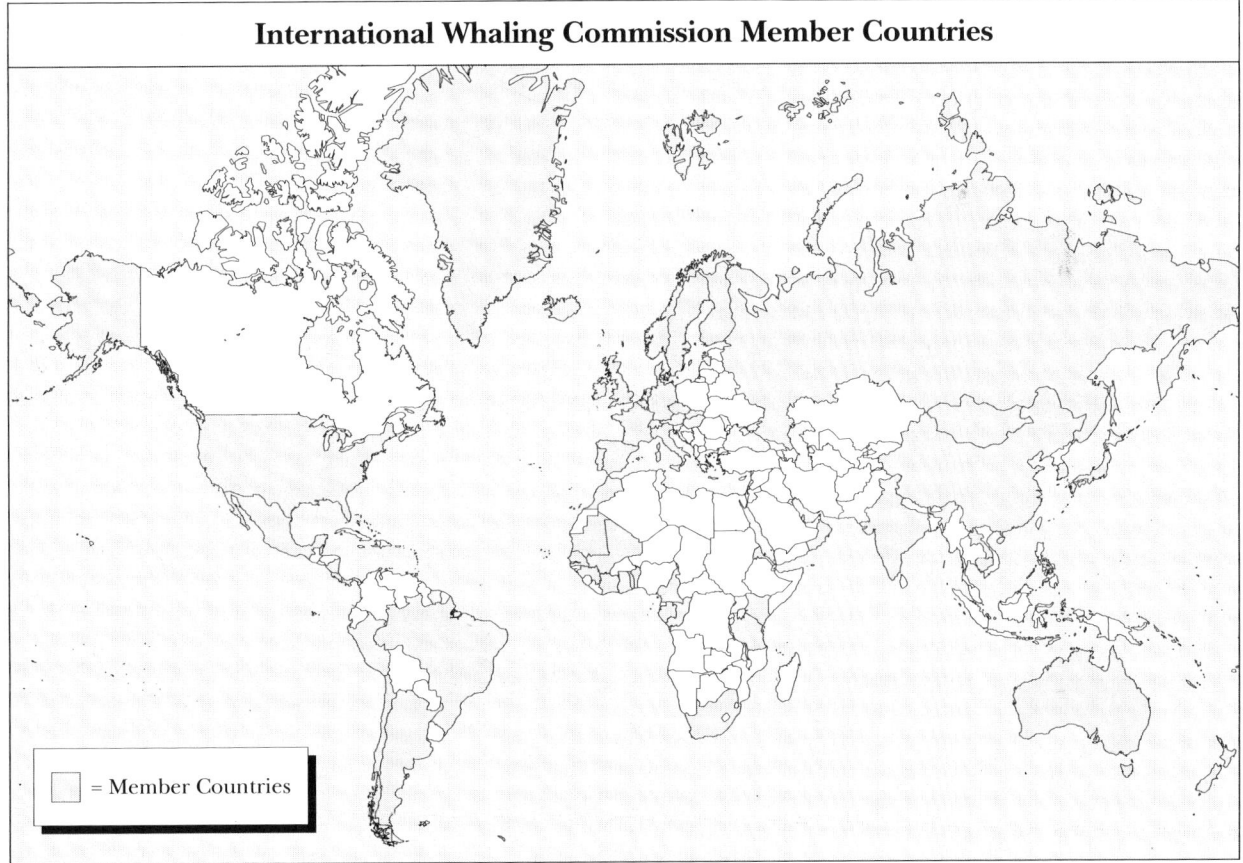

Norway, the Soviet Union, and the United States were among the fifteen charter nations. Japan joined in 1950, and by 1982 membership had climbed to thirty-seven nations; as of 2009, the number was eighty-eight.

The stated mission of the IWC at its establishment was to regulate international whaling to promote the increase of whale stocks, thereby ensuring the continued existence of the lucrative whaling industry. The commission set a whaling season and created sanctuary zones in which whaling was not permitted. Defined in terms of blue whale units (BWUs), the whaling season ended when that year's quota had been met. However, the IWC was virtually powerless to enforce the regulations it implemented, and it allowed member nations a multitude of opportunities to bypass those regulations. As a result, the IWC's activities did little to slow the destruction of whale populations during the first several decades of the commission's existence.

Despite its ineffectiveness, the IWC became an arena of conflict and disagreement among the member states. Because the quota system created a free-for-all environment in which each nation attempted to catch as many whales as possible before the season ended, many nations argued that they did not have an opportunity to catch their fair share of the total number of BWUs available. However, efforts to assign quotas to individual nations proved impossible, as each nation jockeyed for a high percentage of the total at the expense of other nations. Debates grew so acrimonious that the IWC closed its meetings to the public and reporters for a time during the 1960's.

As a regulatory body, the IWC failed to live up to its promise. Quotas were based on estimates of whale populations that were too high. In addition, the commission proved both unwilling and unable to rein in the whaling industries of member nations. In 1982 the IWC voted for an international ban on whaling, which was implemented in 1986. This act reflected the declining economic vitality of the whaling industry more than a concern for the environment or the fate of the whales. As such, the ban came far too late to ensure the survival of many species. After the implementation of the ban, the IWC became more of a forum for nations attempting to circumvent or weaken restrictions on whaling than an organization committed to the preservation of whales.

Thomas Clarkin

FURTHER READING

Andresen, Steinar. "The International Whaling Commission (IWC): More Failure than Success?" In *Environmental Regime Effectiveness: Confronting Theory with Evidence*, by Edward L. Miles et al. Cambridge, Mass.: MIT Press, 2002.

Friedheim, Robert L., ed. *Toward a Sustainable Whaling Regime*. Seattle: University of Washington Press, 2001.

Heazle, Michael. *Scientific Uncertainty and the Politics of Whaling*. Seattle: University of Washington Press, 2006.

Stoett, Peter J. *The International Politics of Whaling*. Vancouver: University of British Columbia Press, 1997.

Land-use policy

CATEGORY: Land and land use
DEFINITION: The way in which a society organizes, plans, and manages social and physical activities on the landscape
SIGNIFICANCE: Policy makers around the world have become increasingly aware of the potential environmental impacts of government and community decisions regarding land use. Since the late twentieth century, approaches to land-use planning have trended toward ecological conservation and preservation.

Human societies are organized to survive in particular environments. The adaptive nature of culture allows a society to respond to changes in the environment and to cause even more changes. Once human beings began to establish cities, where their lifestyles became relatively sedentary, formal land-use policies arose, and decisions about how to manage land became critical to survival. Even during the earliest periods of the Mayan, Egyptian, and other city-state civilizations, land-use control was implemented for religious, agricultural, hunting, and residential purposes. Religious proscriptions dictated the appropriate appearances of structures. Plagues, warfare, and resource distribution demonstrated the need to isolate structures as well as groups of people. Crowding and cultural conflicts made it necessary for groups to create processes to make government decisions about who would get to use particular resources. Land-use policies are even more necessary in the contemporary world of competing interests, growing populations, and diminishing natural resources. Increased scientific knowledge about environmental impacts has fueled the need to ensure that appropriate land-use decisions are made in the interests of survival through long-term resource management.

ROOTS OF LAND-USE POLICY

The earliest land-use policies took the form of religious prohibitions and mandates as the priest class interpreted the needs of the gods for sacred spaces. Kings exercised a divine mandate in interpreting just what could be allowed in sacred areas, while market forces dictated the use of less important secular space. Later, after kings began losing their divine authority, they could still regulate the use of secular space in the name of promoting social order. Social class structure continued to support the allegiance to class-based authority.

After the Renaissance, European societies believed that while God no longer directly intervened in day-to-day activities, his presence was felt in the need to maintain social order through hierarchies. It was believed that this social order must also be reflected in physical space, or in the way in which a landscape was arranged. Accordingly, it was seen as only proper that buildings, towns, and landscapes reflect particular patterns in form and ownership. In the American colonies that became the United States, examples of kingly intervention in land-use decisions can be found in the marking of potential mast pine trees in New England with the king's broad arrow, the designation of village squares, and the reservation of lots for the king's agents. However, the rise of a democratic society provided a shift in decision-making authority to elected representatives.

By the mid-nineteenth century, early planning laws in the United States began to regulate urban tenement housing and prohibit "obnoxious uses." In 1893 the World's Columbian Exposition in Chicago promoted the exchange of planning and design concepts in exhibits by landscape architect Frederick Law Olmsted, artist Saint Gaudens, and others. By 1895 Los Angeles had an ordinance prohibiting the locating of steam shoddying plants (plants that reprocessed scrap wool) within 30.5 meters (100 feet) of a church. In addition to ordinances and the designation of parks, planned communities were initiated. The publication of Ebenezer Howard's *Tomorrow. A Peaceful Path to Real Reform* (1898) launched the garden city movement. In an 1899 decision, the Massachusetts Supreme Court upheld a building height limitation, and by 1909 zones of limits for buildings were upheld in the U.S. Supreme Court.

By the early twentieth century, organized planning efforts were under way in some large cities. Hartford, Connecticut, established a planning board in 1907. By 1909 Wisconsin had passed the first state enabling act for planning, and Los Angeles provided the first American use of zoning to direct future development in a series of multiple zoning ordinances. In 1913 Massachusetts became the first state to make planning mandatory for local governments. Newark hired the first full-time city planner in 1914, and the first comprehensive zoning code in the United States was enacted in New York City in 1916. By 1925 Cincinnati, Ohio, had become the first major U.S. city to adopt a comprehensive plan, and Burlington, Vermont, had authorized a municipal planning commission.

POLICIES AFTER WORLD WAR II

In the 1940's planning for war and postwar public housing brought the U.S. government to review the land-use planning process. The mustering of resources for World War II demonstrated the value of planning, and many towns implemented town plans as postwar prosperity began. In 1962 the Chicago Area Transit Study showed the applicability of cost-benefit studies to planning for suitable development.

In the 1970's Americans became increasingly concerned with the need to control development. Performance standards were upheld in the courts as one mechanism to manage growth through land-use policy. In 1971 the concept of transferable development rights (TDRs) was introduced to help preserve urban landmarks. Creative land-use tools such as conservation easements, controlled access rights, planned unit development (PUD) density credits, and overlay districts were increasingly used in the 1970's as communities expanded their regulatory schemes. In their zeal to manage growth, some communities enacted land-use policies that were discriminatory. In an early decision on discriminatory land-use regulation, in 1975 the New Jersey Supreme Court struck down a restrictive Mount Laurel zoning ordinance on the basis that it did not allow a regional "fair share" of low- and middle-income housing.

By the late 1980's a sufficient number of court cases existed to reduce the likelihood that communities would enact discriminatory ordinances, but a slump in the U.S. economy fueled the challenge of some land-use ordinances and policies as unauthorized "takings"—the erosion or loss of landowner rights through excessive regulation without due compensation. Yet the vast majority of land-use regulations are not generally found to be true takings because at least some economic use of land is allowed. Still, the issue of a taking is frequently raised when regulations deprive an owner of one or more desired uses. The economic climate has a strong influence on the reaction of a population to particular issues such as takings and to land-use regulation in general.

Land-use regulation is generally viewed as the control of two categories: subdivision (physical size or boundary change) and development (physical use or alteration within the boundary). Changing landownership or subdivision is a land use because it affects the management of resources on the land and can fragment habitat. Most land that has been subdivided stays that way; it is unusual for such land to revert to an original larger tract. Land that is used or developed through construction, clearing, or other alteration, including various forms of land management such as agriculture, also undergoes a change to the natural path of succession. In fact, sufficient past alteration through introduction of new species or direct physical action has so altered some landscapes that it is almost impossible to determine their true "natural" condition. It is this perspective that, along with concern regarding environmental impacts from yet-to-occur changes, is most commonly used to justify land-use policies.

Conventional and Conservation Planning

Conventional land-use planning assumes the desirability of economic growth through new development and therefore tends to favor revenue generators. Tools such as zoning are the main techniques for conventional planning. A government land-use plan is implemented through a series of development regulations that differ according to zones. A series of base maps are prepared after the completion of natural, social, and infrastructural resource inventories. The maps can be viewed as both opportunities for and constraints on growth. Individual base maps, when combined with the community's or region's goals and objectives for growth, result in a land-use map containing zones. Each zone reflects a particular category: commercial, industrial, residential, recreational, historical, governmental, agricultural, special, and others that are suggested by the inventory and goal processes. Each category can contain a variety of subcategories based on lot size and range of allowable uses.

The size, configuration, and pattern of the arrangement of lots can all affect growth. For example, large lots reduce the number of houses, while small lots increase the number of houses, reduce the amount of open space, and increase the fragmentation of landownership. The community uses the planning process to agree upon the development capacity of given areas. By using various tools and approaches, a community can seek a balance between achieving appropriate densities and maintaining natural and aesthetic resources; the objective is to achieve sustainability.

Conservation planning is one approach to improving land use within conventional planning. In conservation planning, structures and uses such as septic systems are located away from valued or critical natural resources on a tract. Conventional planning, even if conservation-oriented, can still lead to checkerboard or highway strip patterns of development that are harmful to open space. Clustering is a technique that goes one step further by attempting to preserve or conserve open space by treating it as a natural resource. Other cited benefits of clustering are the fostering of a sense of community, reduction of urban sprawl, and the presentation of traditional village appearances.

In clustering, dwelling units (or commercial structures) are grouped together to allow a larger uninterrupted area of open space (generally 25 to 30 percent), which is often maintained in a residential or commercial subdivision as common land under shared ownership. Some communities encourage clustering through policies that allow a greater density of units or square feet of construction. Thus a 20-hectare (50-acre) tract that might be approved for construction of five houses in a traditional checkerboard subdivision of ten lots might be allowed to contain up to fifteen units of housing if the houses (or condominiums) are clustered and if a specified percentage of the parcel is preserved as open space. The open space might be a separate 8-hectare (20-acre) lot that is deeded to the prospective unit owners as shared or common land to be managed in a certain way.

Conservation planning can be employed on a city- or statewide level through the administration of tiered levels of permits and other development review processes coupled with a system of land-use planning in which designated areas are conserved. An example is Oregon's establishment of greenline boundaries, which limit growth at the edges of cities. Trails and greenbelt corridors also reflect conservation planning, but these require considerable coordination and cooperation when more than one community or state is involved.

Ecological Planning

Ecological land-use planning expands on conventional land-use planning by taking a more integrative perspective of ecosystem dynamics and applying it over a greater period of time than the normal five-year period. To take this comprehensive approach, planners require significant amounts of data about a wide variety of resources and a fairly stable set of goals and objectives in a relatively constant sociopolitical setting. Ecological planning provides the benefit of long-range dynamic planning while attempting to prevent, rather than remedy, problems. However, it is more costly than conventional planning in initial expenditures. The Netherlands and some other northern European countries have a long history of ecological land-use planning, particularly in response to increased population pressures in areas of finite land resources.

The decrease in numbers of rural inhabitants and the growing numbers of suburbanites in the United States in the latter half of the twentieth century caused an increase in the consideration of various planning techniques and tools. Vermont adopted a statewide land-use policy law in which individual development decisions are made by a regional volunteer citizen panel at quasi-judicial hearings in which dispute resolution techniques and consensus building are encouraged. Vermont has found this case-by-case process to work quite well despite the lack of a comprehensive statewide plan and despite an expanding population. Oregon uses a similar process. Florida's Environmental Land and Water Management Act of 1972 allows the state to designate areas of critical interest that local governments must consider when enacting local policies. It also requires state review of development projects that are large or have regional impacts.

The cumulative adverse environmental effects of many small subdivision and construction projects can significantly outweigh the effects of larger projects that are more intensely regulated. By the end of the twentieth century most U.S. states had begun to recognize this and had increased local control of land use. Most had also implemented statewide natural resource programs that reflected ecological understanding of the interactions between changing land use and the need to manage wildlife and other natural resources.

Forms of Land-Use Regulation

Land use can be controlled through incentives or restrictive processes. Incentive-based land-use control approaches include direct funding, grants, tax abatements, trade-offs such as credits for clustering, and other forms of positive feedback. Restrictive land-use control is the form more commonly recognized and employed, notably through the use of permits, licenses, environmental assessments, taxes, and direct prohibitions. Restrictions may be imposed by government or by landowners via covenants or easements. Government restrictive regulations have two forms: prescriptive, in which the objectives and specifications are precisely articulated; and subscriptive, in which the outcomes are specified but the individual means of achieving them are left to the discretion of the owner or community. Critical land uses and issues, such as matters of public health, are more likely to be prescriptively regulated. Less critical matters might be handled by subscriptive means, also called performance-based planning. Even tax structuring can be performance-based when land is taxed based on actual use rather than potential use. This can reduce the pressure for commercial development of high-value or expensive properties.

Most communities and governments use a combination of the two forms of land-use control and the two forms of regulations. Much of land-use policy concerns the manner in which the combination is achieved. Issues of land-use control can become highly politicized, as in the wise-use movement or the controversy involving the northern spotted owl and logging policy in the northwestern United States. In such cases, reaching sufficient community consensus to support consistent regulation can be difficult to achieve because of the differing cultural and social values held by the parties involved.

Although the United States and Canada have contributed greatly to the literature on land-use planning and the development of innovative techniques, North America has relatively weak land-use controls, as does Australia. Northern Europe and countries such as Japan are known for their comprehensive land-use controls In the United States, as in many countries, the redistribution of population and wealth, together with the restructuring of public lands, necessitates constant reexamination of land-use policies and regulations. For developing nations, land-use policies become particularly critical in the evaluation of trade-offs between natural and land-based resources on one hand and economic well-being on the other. Pressures exist for these countries to exploit their natural resources while striving to achieve the prosperity they see in more developed countries such as those in North America and Northern Europe. As global markets have expanded, the need for international dialogue on land-use issues has also grown.

Robert M. Sanford and Hubert B. Stroud

Further Reading

Arendt, Randall. *Rural by Design: Maintaining Small Town Character.* Chicago: American Planning Association, 1994.

Arnold, Craig Anthony. *Fair and Healthy Land Use: Environmental Justice and Planning.* Chicago: American Planning Association, 2007.

Goetz, Stephan J., James S. Shortle, and John C. Bergstrom, eds. *Land Use Problems and Conflicts: Causes, Consequences, and Solutions.* New York: Routledge, 2005.

Johnston, Robert J., and Stephen K. Swallow, eds. *Economics and Contemporary Land Use Policy: Development and Conservation at the Rural-Urban Fringe.* Washington, D.C.: Resources for the Future, 2006.

Marsh, William M. *Landscape Planning: Environmental Applications.* 5th ed. Hoboken, N.J.: John Wiley & Sons, 2010.

Porter, Douglas R. *Managing Growth in America's Communities.* 2d ed. Washington, D.C.: Island Press, 2008.

Sargent, Frederic O., et al. *Rural Environmental Planning for Sustainable Communities.* Washington, D.C.: Island Press, 1991.

League of Conservation Voters

CATEGORIES: Organizations and agencies; activism and advocacy; preservation and wilderness issues
IDENTIFICATION: American nonprofit organization that promotes proenvironmental policies and works to elect political candidates who support such policies
DATE: Founded in 1970
SIGNIFICANCE: The League of Conservation Voters has played an important role in influencing the voting public in the United States with its campaigns for or against candidates for national office based on those persons' support or lack of support for environment-related legislation.

In 1969 environmental leader David Brower created the organization Friends of the Earth (FOE), which later grew to become one of the world's largest networks of grassroots environmental organizations. In 1970, after the first Earth Day, Brower split off some of FOE's staff in Washington, D.C., to establish the League of Conservation Voters (LCV), an organization that would focus specifically on politics, tracking the actions of the U.S. Senate and the House of Representatives and campaigning for proenvironmental candidates and issues. Early issues addressed by LCV included nuclear power, the Trans-Alaska Pipeline, and the protection of endangered species. By the twenty-first century, the group had expanded its range of concerns to include global climate change and clean energy.

Every year since 1970, LCV has issued its National Environmental Scorecard, which rates each member of Congress according to his or her voting record on bills concerning environmental, public health, and energy issues. The Scorecard reports on the results of Senate and House votes on certain bills and gives each member of Congress a percentage score. Representatives from approximately twenty environmental and conservation groups work together to identify which votes should be considered and to determine the percentage score for each member of Congress. Several states have their own branches of LCV, and these track members of their state legislatures in the same way. LCV also issues an annual Presidential Report Card, and during election campaigns it releases information about candidates' environmental records.

Since 1996, LCV has released a list that it calls the Dirty Dozen during election years; this list identifies the twelve sitting members of Congress who have most consistently voted against environmental protection and who are up for reelection. In selecting the legislators who will appear on the list, LCV targets candidates in races that are projected to be close enough that an oppositional campaign could make a difference in the outcome; the organization then actively campaigns against these incumbents, emphasizing their environmental voting records. The candidates selected may be members of the Senate or of the House of Representatives and may belong to any political party, although most have been Republicans. LCV also endorses candidates who take strong proenvironmental stances and supports their campaigns with funding through its LCV Action Fund.

LCV set up the LCV Education Fund in 1985 to support environmental education efforts at the local, state, and national levels. The fund provides small grassroots environmental organizations with training, research assistance, and technology that they could not develop on their own; it also helps local organizations conduct opinion polls and run get-out-the-vote programs.

Cynthia A. Bily

Further Reading

Duffy, Robert J. *The Green Agenda in American Politics: New Strategies for the Twenty-first Century.* Lawrence: University Press of Kansas, 2003.

Gibson, James William. *A Reenchanted World: The Quest for a New Kinship with Nature.* New York: Macmillan, 2009.

League of Conservation Voters. *National Environmental Scorecard '05.* Washington, D.C.: Author, 2006.

Lovins, Amory

CATEGORIES: Activism and advocacy; energy and energy use
IDENTIFICATION: American physicist
BORN: November 13, 1947; Washington, D.C.
SIGNIFICANCE: Lovins, cofounder of the Rocky Mountain Institute, has worked to promote the use of sustainable and clean energy, particularly as a means to attain global stability and security.

A native of Washington, D.C., Amory Lovins was a student at Harvard University and Magdalen College, Oxford, England, during the mid- to late 1960's. He received his master's degree from Oxford University in 1971 and thereafter began working as a consultant physicist specializing in energy concerns and promoting the use of energy sources that are economical, efficient, diverse, sustainable, and environmentally sound. Lovins advocates an approach that maximizes energy efficiency and minimizes environmental impact as the best way to provide for the world's long-term energy needs without incurring the security risks inherent in the use of nuclear power plants or reliance on foreign oil.

Lovins has lectured extensively, has held several visiting academic chairs, and has published hundreds of papers. Among his many books are *World Energy Strategies: Facts, Issues, and Options* (1975), *Soft Energy Paths: Toward a Durable Peace* (1977), and *Non-Nuclear Futures: The Case for an Ethical Energy Strategy* (1980; with John H. Price). Lovins has been the recipient of numerous U.S. honorary doctorates and a 1993 MacArthur Fellowship. He has consulted for utilities, industries, and governments worldwide, has briefed heads of state, and has served on the U.S. Department of Energy's senior advisory board.

In 1979 Lovins married L. Hunter Sheldon, one of the cofounders (and for six years the assistant director) of the urban forestry and environmental education group TreePeople. In 1982 the Lovinses cofounded the Rocky Mountain Institute, a nonprofit resource policy center in Old Snowmass, Colorado, that promotes resource efficiency and global security, with Hunter as the institute's president and executive director and Amory as its vice president and director of research. The institute's stated mission is to "foster the efficient and sustainable use of resources as a path to global security." Its projects focus on such areas as transportation, green design and development, greenhouse gas reduction, water-efficient technologies, economic renewal, and corporate sustainability.

Amory and Hunter Lovins coauthored several books, even after they separated in 1989 (they divorced in 1999), including *Energy/War: Breaking the Nuclear Link* (1980), *Brittle Power: Energy Strategy for National Security* (1982), *Energy Unbound: A Fable for America's Future* (1986; with Seth Zuckerman), *Climate: Making Sense and Making Money* (1997), and *Natural Capitalism: Creating the Next Industrial Revolution* (1999; with Paul Hawken). The couple shared a number of honors, including the 1982 George and Cynthia Mitchell International Prize for Sustainable Development, the 1983 Right Livelihood Award (often called the "alternative Nobel Prize"), the 1989 Delphi Prize (one of the world's top environmental awards, presented by the Onassis Foundation), and the 1993 Nissan Prize at the International Symposium on Automotive Technology and Automation, Europe's largest car technology conference. In a 2009 list published by

Time magazine, Lovins was recognized as one of the world's one hundred most influential people.

Karen N. Kähler

FURTHER READING

Hawken, Paul, Amory B. Lovins, and L. Hunter Lovins. *Natural Capitalism: The Next Industrial Revolution.* New ed. Sterling, Va.: Earthscan, 2005.

Heintzman, Andrew, and Evan Solomon, eds. *Fueling the Future: How the Battle over Energy Is Changing Everything.* Toronto: House of Anansi Press, 2003.

Inslee, Jay, and Bracken Hendricks. *Apollo's Fire: Igniting America's Clean-Energy Economy.* Washington, D.C.: Island Press, 2008.

Maathai, Wangari

CATEGORIES: Activism and advocacy; preservation and wilderness issues

IDENTIFICATION: Kenyan environmentalist and social activist

BORN: April 1, 1940; Ihithe village, Nyeri District, Kenya

SIGNIFICANCE: A visionary and activist in the fight against deforestation in Africa and beyond, Maathai has spearheaded various initiatives that have resulted in the planting of billions of trees and have brought global attention to this critical environmental issue.

Even before she won the Nobel Peace Prize in 2004, Wangari Maathai was well known for the positive impacts of her efforts to raise environmental awareness, particularly in regard to the issue of deforestation. Born Wangari Muta, Maathai grew up in Kenya as the daughter of subsistence farmers. She received scholarships that enabled her to earn bachelor's and master's degrees in biological science in the United States, and she also spent some time studying in Germany.

Upon her return to Kenya, Maathai studied anatomy at the University of Nairobi, but her interests turned to the intertwined issues of poverty and deforestation, partly because she realized how much the practice of deforestation had changed Kenya's landscape in her absence. In 1969 she married Mwangi Mathai, with whom she would eventually have three children, and in 1971 she became the first woman in central and eastern Africa to earn a Ph.D. Her husband became politically active, and in 1974 he campaigned for a seat in the Kenyan parliament. In part to help him keep a campaign promise to create new jobs, Maathai founded a business called Envirocare, which paid people to raise tree seedlings in nurseries for eventual transplantation across Kenya. Maathai's progressive political views strained her marriage, however, and Mwangi Mathai eventually sued for divorce. When the marriage ended, he asserted his legal right to demand that she change her surname, and, as she relates in her memoir, *Unbowed* (2006), instead of changing her surname completely Maathai instead chose to insert an extra "a."

Maathai's personal difficulties did not lessen her desire to have a positive influence on environmental outreach and activism. In 1977 Maathai renamed Envirocare the Green Belt Movement and gained the support of Kenya's National Council of Women. The organization adopted the motto of "One person, one tree," which led to the goal of planting fifteen million trees, one for each person in Kenya. The group far exceeded the goal, planting more than twice that number of trees by the early twenty-first century.

At the same time, Maathai expanded her activities into other areas of the environmental movement. She successfully campaigned against the building of a planned skyscraper in Nairobi's Uhuru Park and opposed the government's attempts to sell valuable forestland to developers. Her work met with much opposition and she was imprisoned several times, but her international influence increased to the point that it became difficult for the authorities to detain her without cause.

In 2002 Maathai was elected to Kenya's parliament, and she was appointed the following year to the post of assistant minister of the environment, natural resources, and wildlife. Her tireless work was recognized worldwide when she became the first African woman to win the Nobel Peace Prize in 2004. Maathai then helped the United Nations Environment Programme launch the Billion Tree Campaign. The campaign's initial goal of planting one billion trees was reached more quickly than expected, and a new goal of planting more than seven billion trees by the end of 2009 was also exceeded. The significance of Maathai's contribution and inspiration to this campaign, which was carried out by workers and volunteers in more than 170 counties, is clear.

Amy Sisson

FURTHER READING

Maathai, Wangari. *The Green Belt Movement: Sharing the Approach and the Experience*. Rev. ed. New York: Lantern Books, 2003.

_____. *Unbowed: A Memoir*. New York: Alfred A. Knopf, 2006.

McToxics Campaign

CATEGORIES: Activism and advocacy; waste and waste management

THE EVENT: Consumer boycott against McDonald's restaurants

DATES: 1989-1990

SIGNIFICANCE: The McToxics Campaign prompted the huge McDonald's fast-food restaurant chain to replace the polystyrene food containers it had been using with more environmentally friendly packaging.

In 1989, the Environmental Defense Fund (EDF) inaugurated a grassroots campaign against McDonald's restaurants. By convincing consumers to boycott McDonald's, EDF hoped to persuade the fast-food giant to stop packaging its food in polystyrene "clamshells," which were cheap and efficient food insulators but did not readily biodegrade and were suspected of emitting chlorofluorocarbons (CFCs) harmful to the earth's ozone layer. EDF members worked to educate children about the environmental consequences of polystyrene packaging, then employed the young activists in their protests. Ronald McDonald, the company's well-known advertising character designed to appeal to young consumers, was redubbed "Ronald McToxic" by the campaign's directors.

Predictably, the children's participation drew media attention, and the protest took effect as consumers began to boycott McDonald's. Some wrote angry letters; others mailed empty clamshells to the company's offices. As the corporation's profits dropped, McDonald's lawyers began urging the company to undertake the packaging changes demanded by EDF and its supporters. On August 1, 1990, McDonald's came to a tentative agreement with EDF, and on November 1, 1990, McDonald's announced its decision to abandon the use of polystyrene packaging.

Nevertheless, the agreement was not without detractors. Some scientists expressed doubts that the changeover from polystyrene to paper packaging would produce environmental benefits, questioning whether the use of clamshells represented a legitimate environmental hazard. In addition, some environmental extremists objected to EDF's accommodation of corporate culture. Nevertheless, most observers agreed that the McToxics campaign and its conclusion represented a welcome example of how environmental and economic interests could be made to coexist.

Alexander Scott

Marine Mammal Protection Act

CATEGORIES: Treaties, laws, and court cases; animals and endangered species

THE LAW: U.S. federal legislation intended to provide for safe environments for marine mammals

DATE: Enacted on October 21, 1972

SIGNIFICANCE: The provisions of the Marine Mammal Protection Act contributed to improvements in the populations of various species, including seals, sea otters, and whales, but debates continue regarding the law's enforcement and its effects.

The Marine Mammal Protection Act (MMPA) prohibits ownership or importation of any marine mammal or any products of marine mammals. However, it does allow a limited catch by Alaska Natives and Native Americans for purposes of material survival or for reasons related to cultural heritage. The act was amended in 1994 to restructure jurisdiction and enforcement of the provisions of the law and to establish guidelines for transportation of marine mammals. Controversies exist among fishing interests, environmentalists, and members of indigenous cultures as to the interpretation and effects of the MMPA.

When the MMPA became law in 1972, the U.S. Fish and Wildlife Service (FWS), which is part of the Department of the Interior, became responsible for manatees, dugongs, polar bears, walruses, and sea otters. The National Marine Fisheries Service (NMFS), a division of the Department of Commerce, was assigned management of whales, dolphins, sea lions, fur seals, elephant seals, monk seals, true northern seals, and southern fur seals. The 1994 amendments stipulated stronger fishing regulations, especially the use of improved equipment to reduce the number of accidental killings of marine mammals and to exclude the bycatch of turtles, nontarget fish, and undersized fish of the targeted species.

Before the amendments, jurisdiction over care and transport of captive marine mammals was shared by the NMFS and the Department of Agriculture's Animal and Plant Health Inspection Service (APHIS). The amendments eliminated the NMFS's part of the administration and enforcement, which caused concern among many environmentalists because APHIS officials are not as experienced in working with marine mammals as are personnel of the NMFS. The Humane Society of the United States appealed for reinstatement of the NMFS as a joint authority, but APHIS was delegated sole authority. Zoos and aquariums supported APHIS control, as that agency made it easier for these institutions to capture and transport marine mammals. APHIS required only that public facilities that already owned any marine mammals send notification to APHIS after acquiring additional mammals, whereas previously the institutions had been required to obtain permits before acquiring such animals.

Other changes brought about by the amendments eased regulations on scientists and researchers, who no longer are required to obtain permits to conduct studies of marine mammals, unless their work has the potential to harm the animals. Many ecologists reacted positively to the 1994 amendments to the MMPA because the amendments emphasized the importance of maintaining healthy ecosystems, particularly in the waters off the northwestern and northeastern coasts of the United States, where the seal and sea lion populations had been declining at an alarming rate.

The MMPA made illegal many of the human activities that threatened marine mammal populations in the past. Sea otters had been overhunted for their skins, but government protection has allowed their populations to recover around Prince William Sound and off the California coast. Whales are protected by the MMPA and by the ban on whaling instituted by the International Whaling Commission. However, the natural renewable resources on which whales feed may be endangered, as some countries are harvesting large quantities of krill, which is the mainstay of many whales' diets and an important link in the marine food chain.

As human populations increase worldwide and become more industrialized, demands on the oceans as a food source and a place to dump chemicals also increase, as does noise pollution from sonic testing and boat traffic. Continued publicity and pressure from environmental advocacy groups such as Greenpeace and the World Wide Fund for Nature have led wildlife managers, fishers, animal rights supporters, and scientists to work together under the terms of the MMPA as amended in 1994.

Dale F. Burnside with Aubyn C. Burnside

FURTHER READING
Curnutt, Jordan. *Animals and the Law: A Sourcebook.* Santa Barbara, Calif.: ABC-CLIO, 2001.
Ray, G. Carleton, and Jerry McCormick-Ray. *Coastal-Marine Conservation: Science and Policy.* Malden, Mass.: Blackwell, 2004.

Marshall, Robert

CATEGORIES: Activism and advocacy; forests and plants; preservation and wilderness issues
IDENTIFICATION: American forester and plant physiologist
BORN: January 2, 1901; New York, New York
DIED: November 11, 1939; on a train en route from New York to Washington, D.C.
SIGNIFICANCE: Marshall influenced both government policy and public opinion through his numerous writings on the need for wilderness conservation and through his participation in the Wilderness Society, an organization he cofounded.

Robert Marshall was born in 1901. He was the son of Florence Lowenstein Marshall and constitutional lawyer and conservationist Louis Marshall, who had been a delegate at the New York State Constitutional Convention of 1894, which placed in the state constitution the famous provision that the New York Forest Preserve shall be "kept forever wild." Marshall's extensive utilization of the family library introduced him to books and topographical surveys of the Adirondack Mountains. At the age of fourteen, he, along with his brother George and a guide, ascended a high Adirondack peak, thus cementing a lifelong love affair with wilderness exploration and celebration.

Marshall's higher education at three universities (bachelor of science in forestry at Syracuse, master's degree in forestry at Harvard, and Ph.D. in plant pathology at The Johns Hopkins University) served him well as he developed a literary and resource management career. During his early professional work in the U.S. Forest Service, he had the opportunity to observe large, unbroken wilderness conditions, which served as the catalyst and foundation for his first major article on extensive natural landscapes. Titled "The Problem of the Wilderness" and appearing in the February, 1930, issue of *The Scientific Monthly*, it was a clarion call for setting aside and protecting large tracts of land in their natural and, to the extent possible, primeval condition. The article elucidated four salient wilderness themes: its great beauty and wildness, with integrated aesthetic, mental, and physical values; the rapid disappearance of wilderness; the need for human beings to look beyond commodity value of resources in wilderness as the sole arbitrator of its value; and the urgency to act for wilderness preservation.

In 1931 Marshall settled in Washington, D.C., and immediately devoted his efforts to writing assignments. Collaborating with the U.S. Forest Service on a book titled *A National Plan for American Forestry* (1932), he contributed sections on national parks, wilderness, and recreation. One year later he published *The People's Forests*, in which he articulated the importance of conserving water, soil, and forests. Again he upheld the need to preserve forested areas through arguments related to human aesthetic needs, arguing that such preservation is of pivotal importance to contemporary society.

In 1933 Marshall was appointed director of forestry at the Office of Indian Affairs, where he helped develop sixteen wilderness areas on Indian reservations. Two years later, he was a leader of eight people who founded the Wilderness Society. In 1937 he became chief of the new Forest Service Division of Recreation and Lands and immediately began moving official Forest Service policy toward supporting wilderness. He also drafted new administrative regulations relating to a classification system for wilderness and wild areas. Approval for these regulations came just months before Marshall's untimely death in 1939 at age thirty-eight. In 1964, as the Wilderness Act became law, the twentieth area to be named to the National Wilderness Preservation System was the Bob Marshall Wilderness Area in the Flathead and Lewis and Clark national forests.

Charles Mortensen

FURTHER READING

Herron, John P. *Science and the Social Good: Nature, Culture, and Community, 1865-1965*. New York: Oxford University Press, 2010.

Marshall, Robert. *Alaska Wilderness: Exploring the Central Brooks Range*. Edited by George Marshall. 3d ed. Berkeley: University of California Press, 2005.

_____. *The People's Forests*. 1933. Reprint. Iowa City: University of Iowa Press, 2002.

Monkeywrenching

CATEGORIES: Activism and advocacy; philosophy and ethics

DEFINITION: Direct-action tactics used by radical environmentalists to disrupt activities they believe degrade the environment

SIGNIFICANCE: The activities known as monkeywrenching are the source of considerable controversy. Many environmentalists see such radical tactics as damaging to their work to gain the support of the public for government policies and regulations aimed at preserving and protecting the environment.

The term "monkeywrenching" was first used in 1904 to refer to the sabotaging of factory machinery by throwing a monkey wrench, or a spanner with a movable jaw, into the works. Such acts were preceded in early nineteenth century England by the actions of Luddites and machine breakers who protested the mechanization of workplaces during the Industrial Revolution. The term was appropriated later by environmentalist and author Edward Abbey in his 1975 novel *The Monkey Wrench Gang*, which details the exploits of three men and one woman who take the law into their own hands to defend the wilderness from excavation. They perform acts of sabotage, or ecotage, on road-building equipment and entertain visions of blowing up Arizona's Glen Canyon Dam.

The radical environmental group Earth First!, founded in 1980, has used civil disobedience and monkeywrenching tactics to defend wilderness areas from developers. The general idea behind monkeywrenching is to create a stir, delay or halt projects, and gain publicity for the cause. Dave Foreman, one of the founders of Earth First!, published *Ecodefense: A Field Guide to Monkeywrenching* in 1985 and *Confessions of an*

Eco-Warrior in 1991. He credits Abbey's book as a major motivation and inspiration. In 1990 Earth First! promoted the Redwood Summer, a ten-week campaign to slow the logging of redwoods. Among the tactics used during the campaign was tree spiking, in which large, long nails are driven into trees to dissuade loggers. Because the nails can potentially shatter chain saws and thus hurt loggers, Foreman, in *Ecodefense*, advised monkeywrenchers to warn loggers when an area has been spiked. Incidents have occurred, however, in which no notice of spiking was given.

Another tactic used by monkeywrenchers is tree sitting, in which activists physically occupy trees to prevent their being cut down. In 1998 a California woman lived for many months in a tree to prevent logging of a particular grove. Earth First! activists have also been known to knock down billboards and to destroy heavy equipment used for land clearing and development. The self-proclaimed "navy of Earth First!'s army" is the Sea Shepherd Conservation Society, directed by Paul Watson, who published *Ocean Warrior* in 1994. This group uses direct action to prevent whaling ships and others from killing and capturing marine mammals.

Monkeywrenchers tend to be determined and not easily reformed by the experience of incarceration; many activists have been detained multiple times. The ideological viewpoint of those supporting monkeywrenching is strongly preservationist. Although some are influenced by the writings of such early preservationists as John Muir and Henry David Thoreau, they are far more militant in their activities.

The theory behind monkeywrenching stems from a field of thought known as deep ecology, the core tenet of which is biocentrism, a belief that the human species is just one member of a biological community in which all species have equal standing. This view places humans on the same level as every other living thing. Followers of this philosophy feel that human conduct should proceed from an understanding that all forms of life, no matter how big or small, have an equal right to exist. Therefore, they claim to fight for organisms that cannot defend themselves against industrial intrusion.

Oliver B. Pollak and Aaron S. Pollak

FURTHER READING

Foreman, Dave, and Bill Haywood, eds. *Ecodefense: A Field Guide to Monkeywrenching*. 3d ed. Chico, Calif.: Abbzug Press, 2002.

Scarce, Rik. *Eco-Warriors: Understanding the Radical Environmental Movement*. Updated ed. Walnut Creek, Calif.: Left Coast Press, 2006.

Muir, John

CATEGORIES: Activism and advocacy; preservation and wilderness issues
IDENTIFICATION: Scottish American naturalist, preservationist, and writer
BORN: April 21, 1838; Dunbar, Scotland
DIED: December 24, 1914; Los Angeles, California
SIGNIFICANCE: Muir, one of America's most notable preservationists and a founder of the Sierra Club, introduced Americans to California's Sierra Nevada and worked hard to protect much of the region's wilderness, including Yosemite, against development.

Born in Scotland, John Muir emigrated to the frontier area of Wisconsin in 1849 with his family. Muir's father was a dominant but negative influence

Naturalist and preservationist John Muir. (Library of Congress)

on his life. The elder Muir was deeply religious but viewed the Christian god as one of justice rather than of love. Muir turned away from his father's repressive religion, substituting for it an intense love for nature; for Muir, the divine seemed manifest in the wilderness.

While laboring on the family farm, Muir became an observer of the environment. He was also something of an inventor and worked as a craftsman. After attending the University of Wisconsin, from which he did not graduate, Muir trekked through the southern United States in 1867 hoping to journey on to South America. Instead, he boarded a ship in New York that was bound for California; he arrived in San Francisco in 1868. Muir had no love for cities; he could cross a mountainous wilderness without maps, but he was lost in any urban area. To him, cities were mentally and morally corrupting. He once wrote, "There is not a perfectly sane man in San Francisco."

Muir spent his first summer in California in the Central Valley; in the spring of 1869 he journeyed into the Sierra Nevada as supervisor of a sheep herd. That year was an epiphany for Muir. Yosemite Valley was still almost virginal, and the surrounding cliffs and mountains drew him on as nothing before had done. He stayed in Yosemite the following winter, working in a lumber mill and escaping into the wilderness whenever possible. He wrote of the experience, "I am bewitched, enchanted. . . . I have run wild." With only bread, tea, and a blanket, Muir explored much of the central Sierras.

In 1870 the philosopher and essayist Ralph Waldo Emerson visited Yosemite, and Muir eagerly sought him out. Parlor Transcendentalist Emerson and natural man Muir had much in common, but Muir was unable to convince Emerson to camp under the trees and stars. Muir's attachment to the Sierras was not merely religious, philosophical, or emotional. He was a keen natural scientist. One of his major scientific contributions was his theory that glaciation is the key to explaining the existence of Yosemite Valley and the other canyons and mountains of the Sierras. This idea contravened the accepted doctrine that Yosemite Valley had been formed by a gigantic catastrophe that caused the valley floor to fall several thousand feet. Muir's beliefs were initially dismissed, but in the high country he discovered residual glaciers, validating his claim. In 1879 Muir became one of the first people to explore Alaska's Glacier Bay.

Something of a loner and an eccentric, Muir had few close friends. One of these was Jeanne Carr, whose husband, Ezra, was a professor of geology and chemistry at Wisconsin who later moved to Berkeley, California. Jeanne was very supportive of Muir and his ventures, and it was she who convinced him to write about his experiences, in articles published first in the *Overland Monthly* and later in New York's *Harper's* and *Scribner's* magazines. Muir labored with his writing, but his articles and books are marvelously descriptive, with a visual immediacy that has kept his work in print for many decades. It was through these articles that Muir and his beloved Sierras became known to many Americans.

Muir was always fearful of the impact of civilization on the wilderness. The fragile alpine environments were unsuited to heavy grazing by sheep and cattle, and human visitors to Yosemite and beyond often wished to bring all of their civilized amenities with them. One of Muir's greatest successes was in preserving Yosemite, then under state control and in danger of overdevelopment. With the support of *Century* magazine's Robert Underwood Johnson, Muir mounted a successful campaign that saw Yosemite Valley and much of the surrounding territory become a national park in 1890.

Muir's greatest failure, however, also concerned Yosemite. San Francisco's water system was antiquated and inadequate, and local officials saw a solution in tapping the waters of the Tuolumne River, which ran through Hetch Hetchy Valley; Muir had long praised the beauty of the valley, which, at his urging, had been included within the boundaries of Yosemite park. The issue of damming Hetch Hetchy became a national conflict in the early twentieth century. Some noted conservationists favored it, most notably Gifford Pinchot, chief forester under U.S. president Theodore Roosevelt and, briefly, William Howard Taft. Conservationists such as Pinchot believed that nature and its resources should be made available for human use. Muir, on the other hand, was a preservationist who believed that nature has a right to exist without human interference. Muir and his supporters fought the proposal for years but finally lost in 1913. Soon Hetch Hetchy Valley disappeared in the rising waters when the river was dammed.

Muir died one year later. He left a stirring legacy, not least in the Sierra Club, which he helped found in 1892 and for which he served as president until the end of his life. Many later conservationists and environmentalists have been inspired by Muir's preservationist ethic. No other figure in the environmentalism movement has been so widely honored; California alone boasts many sites named for Muir, including the Muir Woods near San Francisco, Muir Grove in Sequoia National Park, and the Sierras' John Muir Trail.

Eugene Larson

FURTHER READING

Cohen, Michael P. *The History of the Sierra Club, 1892-1970.* San Francisco: Sierra Club Books, 1988.

Fox, Stephen. *John Muir and His Legacy: The American Conservation Movement.* Boston: Little, Brown, 1981.

Miller, Rod. *John Muir: Magnificent Tramp.* New York: Forge, 2005.

Muir, John. *Essential Muir.* Edited by Fred D. White. Berkeley, Calif.: Heyday Books, 2006.

Perrottet, Tony. "John Muir's Yosemite: The Father of the Conservation Movement Found His Calling on a Visit to the California Wilderness." *Smithsonian* 39, no. 4 (July, 2008): 48-55.

Turner, Frederick. *John Muir: Rediscovering America.* 1985. Reprint. Cambridge, Mass.: Perseus, 2000.

Nader, Ralph

CATEGORIES: Activism and advocacy; human health and the environment
IDENTIFICATION: American political activist and consumer advocate
BORN: February 27, 1934; Winsted, Connecticut
SIGNIFICANCE: As a consumer advocate, Nader has been a champion of the underdog—including the poor, the elderly, and members of minority groups—against corporate and political power structures in the United States.

Ralph Nader's interest in environmental issues began long before the environmentalist movement emerged in the late 1960's. When Nader was a child, his parents instilled in him a respect for the duties of citizenship in an industrial democracy. While a student at Princeton University, Nader fought to prevent the spraying of trees on the campus with the pesticide dichloro-diphenyl-trichloroethane (DDT). At Harvard, where he attended law school, he was an editor of the *Harvard Law Record*, which he attempted to change from a dry academic publication to a forum for ideas on social reform.

Nader took his energy and vision for change to Washington, D.C., in the mid-1960's. He was hired by the U.S. Department of Labor's policy planning division as a consultant on highway safety. This job enabled him to amass extensive information on the automobile industry, which he used in writing *Unsafe at Any Speed* (1965), a blatant indictment of the Detroit automobile industry, specifically General Motors. The publication of this book moved Nader to center stage as a consumer advocate.

Beginning in 1969, Nader founded a number of public interest organizations, including the Center for Study of Responsive Law, the Corporate Accountability Research Group, the Public Interest Research Group, and the Clean Water Action Project. These groups focus on a wide range of consumer issues and give high priority to such environmental concerns as safety standards in the workplace, clean and renewable energy resources, automobile emissions standards, clean water standards, and food safety. Nader and his organizations were instrumental in securing passage of major environmental legislation during the 1970's. Nader also played a key role in the establishment of the U.S. Environmental Protection Agency, the Occupational Safety and Health Administration, and the Consumer Product Safety Commission.

Even though the consumer rights and environmentalist movements lost some momentum in the 1980's, Nader's efforts for reform did not decline. The organizations he founded drew supporters who feared that President Ronald Reagan's administration would undo earlier legislative victories. Nader adapted his agenda to address new threats to consumers, such as proposed limits on liability for corporate negligence, restrictions on regulatory agencies protecting consumers, and trade agreements threatening workers and weakening environmental standards. In 1996 Nader ran for president of the United States on the Green Party ticket; this move gave him a new forum from which to articulate his environmental agenda and expose corporate and political leaders' priorities on protecting their narrow interests over the public welfare. He ran for president as the Green Party's candidate again in 2000, and he campaigned for the office as an independent in 2004 and 2008.

Nader has coauthored and coedited several books on environmental topics, including *The Menace of Atomic Energy* (1977), with John Abbotts, and *Who's Poisoning America* (1981), with Ronald Brownstein and

John Richard. He published the autobiographical *Crashing the Party: Taking on the Corporate Government in an Age of Surrender* in 2002 and the memoir *The Seventeen Traditions* in 2007.

Ruth Bamberger

FURTHER READING

Carter, Neil. "Party Politics and the Environment." In *The Politics of the Environment: Ideas, Activism, Policy*. 2d ed. New York: Cambridge University Press, 2007.

Marcello, Patricia Cronin. *Ralph Nader: A Biography*. Westport, Conn.: Greenwood Press, 2004.

Nader, Ralph, Ronald Brownstein, and John Richard, eds. *Who's Poisoning America: Corporate Polluters and Their Victims in the Chemical Age*. San Francisco: Sierra Club Books, 1981.

National Audubon Society

CATEGORIES: Organizations and agencies; activism and advocacy; animals and endangered species

IDENTIFICATION: American nonprofit organization devoted to the conservation of birds and other wildlife

DATE: Established in 1905

SIGNIFICANCE: The National Audubon Society has been successful in publicizing environmental issues relating to wildlife preservation, in lobbying for and supporting federal and state conservation legislation, and in disseminating greater knowledge of wildlife species and their natural environments through its publications.

The National Audubon Society derives its name from John James Audubon, the noted French American painter who was author and illustrator of *The Birds of America* (1838). The earliest Audubon Society was founded in 1886 by George Bird Grinnell, a pioneering conservationist who grew up with the Audubon family. Though it did not last, becoming defunct in 1888, this society spawned several state Audubon societies, the first of which was the Massachusetts Audubon Society founded by Harriet Hemenway in 1896. Primarily through the efforts of ornithologist William Dutcher, these groups affiliated into the National Association of Audubon Societies in 1905 (with Dutcher as first president), though each state society was, and remains, independent, with its own particular emphasis and program agenda, depending on prevailing local conditions. In 1940, the National Association became the National Audubon Society.

Even before the incorporation, the societies were successful in pushing for legislation to curtail or prohibit the killing of waterfowl for the purpose of using the birds' plumage in clothing. (In honor of that success, the National Audubon Society adopted an image of the great egret as the organization's logo emblem in 1953.) In 1918, the National Association achieved a major victory through its role in the passage of the Migratory Bird Treaty Act.

Although the society's purpose as originally conceived was to raise awareness of, and forestall, the potential extinction of bird species, its mission has expanded over the years to include a more general and overarching protection of the environment as a whole and, in particular, the plant and animal species inhabiting it—without abandoning its pronounced emphasis on birdlife. Wildlife sanctuaries have long been a cornerstone of Audubon Society activity, and the national organization and individual state societies operate many such refuges, including Pelican Island and Corkscrew Swamp in Florida, Paul J. Rainey Sanctuary in Louisiana, Hog Island in Maine, Tennille Creek in Oregon, and the Lillian Annette Rowe Bird Sanctuary in Nebraska.

In 1934 Roger Tory Peterson became the society's educational director and also published *A Field Guide to the Birds*, the first nature field guide. Subsequently, the publication of a series of pocket guides covering all aspects of nature developed into a major National Audubon Society initiative. In addition to such guides and other publications, the society publishes the bimonthly magazine *Audubon*, which began as *Bird-Lore* in 1899. An important element in the society's success and continued influence is its focus on the popular, amateur, and local/community levels, particularly through its sponsorship of significant birding (birdwatching) activities.

Raymond Pierre Hylton

FURTHER READING

Anderson, John M. *Wildlife Sanctuaries and the Audubon Society: Places to Hide and Seek*. Austin: University of Texas Press, 2000.

Graham, Frank, Jr., with Buchheister, Carl W. *The Audubon Ark: A History of the National Audubon Society*. New York: Alfred A. Knopf, 1990.

Obmascik, Mark. *The Big Year: A Tale of Man, Nature, and Fowl Obsession.* New York: Free Press, 2004.

Rhodes, Richard. *John James Audubon: The Making of an American.* New York: Alfred A. Knopf, 2004.

Natural Resources Defense Council

CATEGORIES: Organizations and agencies; activism and advocacy

IDENTIFICATION: American nonprofit organization dedicated to the protection of natural resources

DATE: Established in 1970

SIGNIFICANCE: The main goals of the Natural Resources Defense Council are to protect the environment from urban sprawl, reduce pollution, prevent habitat destruction, promote actions to mitigate global warming, and increase the use of renewable energy. The organization has fought successfully to protect natural resources and is considered one of the most influential environmental organizations in the United States.

In the 1960's, when the environmental movement was starting to gain traction in the United States, a group of law students and attorneys under the leadership of John Adams created the Natural Resources Defense Council (NRDC). A nonprofit and nonpartisan environmental organization based in New York City, it was founded with a $400,000 grant from the Ford Foundation. Its stated mission is to safeguard the earth, its people, plants, and animals, and the natural systems on which all life depends. The primary strategy of NRDC is to lobby the U.S. Congress and government agencies to promote the implementation of public policies targeting the sound and sustainable use of natural resources. With offices in Washington, D.C., Los Angeles, Chicago, New York, San Francisco, and Beijing, China, and membership above 1.3 million, NRDC is considered one of the most powerful environmental groups in the United States.

NRDC works on a broad range of environmental issues through a number of different programs. Its air and energy program focuses on issues such as air pollution, global warming, and the development of energy-efficient, renewable energy sources. Its waters and oceans program promotes the protection of fish populations, water quality, wetlands, and oceans. Its health program addresses such environmental problems as the contamination of drinking-water supplies by harmful chemicals, while its urban program focuses on environmental justice and the particular problems seen in urban environments, such as poor air and water quality. NRDC's land program strives to promote improvement in such areas as private and national forest management. Its nuclear program keeps track of new developments regarding nuclear weapons and nuclear power, and its international program addresses worldwide environmental issues such as the protection of rain forests and biodiversity, the preservation of wildlife and natural habitats, and the protection of marine life and oceans. NRDC also carries out public education programs and sponsors scientific research projects.

NRDC is a member of the Nuclear Weapons Complex Consolidation Policy Group, which aims to convince lawmakers around the world of the need to reduce nuclear weapons and to explore sustainable sources of energy other than nuclear power. Thus NRDC's position on nuclear energy is that nuclear power is not the solution to weaning the United States from its dependence on foreign oil, because nuclear power generation represents a danger to humans and the environment.

Lakhdar Boukerrou

FURTHER READING

Bevington, Douglas. *The Rebirth of Environmentalism: Grassroots Activism from the Spotted Owl to the Polar Bear.* Washington, D.C.: Island Press, 2009.

Corbett, Julia B. *Communicating Nature: How We Create and Understand Environmental Messages.* Washington, D.C.: Island Press, 2006.

Miller, G. Tyler, Jr., and Scott Spoolman. "Politics, Environment, and Sustainability." In *Living in the Environment: Principles, Connections, and Solutions.* 16th ed. Belmont, Calif.: Brooks/Cole, 2009.

Nature writing

CATEGORIES: Activism and advocacy; philosophy and ethics

DEFINITION: Nonfiction writing about the natural environment

SIGNIFICANCE: Since the nineteenth century, nature writers have been highly influential contributors to the environmental movement, often raising public awareness of such issues as the dangers to human health of pollution, the negative impacts of the loss of wilderness and of animal species, and the potential harms of various human activities for the long-term health of the planet.

Nature writing celebrates wilderness while simultaneously discouraging environmental exploitation. The field of nature writing includes environmental writing, environmental journalism, and ecocriticism (or ecological criticism), an interdisciplinary study of the environment and literature combining all the sciences to develop solutions for environmental problems.

Since the first Earth Day in 1970, nature writing has increased dramatically in popularity. Nature writing is heavily based on scientific information, research, and facts about the natural world, but the style in which it is often presented provides a unique and broad perspective that reaches a diverse audience. Nature writers frequently write in the first person, a feature that contributes to the emotional and inspirational tone of their work.

Personal observations and philosophical reflections on nature and the environment make up a large portion of nature writing. Among the many types of nature writing, the most basic form is writing that simply conveys information about the natural world, such as through a field guide or other factual or natural history reporting. Natural history essays include important facts about nature and the environment but also incorporate literary elements and meaning or interpretation. Another form of nature writing is based on the personal experiences of the author in nature and tends to be very emotional and personal. Some forms of nature writing include philosophical interpretations of nature and tend to be abstract and scholarly in tone.

Role in Environmental Awareness

For centuries, writers have explored connections between human beings and the natural world. Early nature writers focused on the environment as secondary to humanity's needs and human progress, but by the middle of the nineteenth century this began to change. Nature writers such as Henry David Thoreau and Ralph Waldo Emerson began to reinterpret the significance of nature and the relationship of humans to nature, initiating a movement that led to a dramatic evolution in environmental thought and ethics. The viewpoint shifted toward one in which the environment was recognized as more than simply a natural resource.

The writings of Thoreau, for example, increased public awareness that the natural environment had considerably more value to human beings than simply the natural resources it offered for exploitation; Thoreau emphasized that nature and the environment could be sources of spiritual truth and support. Together with other prominent nature writers such as John Muir and Aldo Leopold, Thoreau, Emerson, and others served to educate the public about nature and the environment.

Prominent Nature Writers

The beginning of modern nature writing can be traced to natural history works popular in the second half of the eighteenth century through the nineteenth century, including writings by Gilbert White, William Bartram, John James Audubon, and Charles Darwin. Other explorers, collectors, and naturalists also contributed to these collections. Thoreau, who wrote many volumes on natural history and the environment, is often considered the father of American nature writing, although his writings were preceded by those of John Bartram and his son William. Many environmental historians consider Thoreau's *Walden: Or, Life in the Woods*, his 1854 memoir, as the beginning of a critical movement in environmental thinking and writing. In subsequent years, this movement flourished and included contributions by John Burroughs, Ralph Waldo Emerson, John Muir, Aldo Leopold, Rachel Carson, and Edward Abbey. Emerson was considered a nature essayist; his writings in the late nineteenth century inspired the works of John Muir, who wrote about his personal experiences in nature, particularly in California's Sierra Nevada.

Rachel Carson's 1962 book *Silent Spring* was critical in raising environmental awareness and fueling public concern about pollution. The book documented the environmental impacts of the uncontrolled spraying of the insecticide dichloro-diphenyl-trichloroethane (DDT) and questioned the rationale that led to the release of substantial amounts of chemicals into the environment when their potential ecological and health impacts were unknown. *Silent Spring* fueled the modern environmental movement by exposing the dangers of indiscriminately used pesticides and fertilizers. It catalyzed a new way of looking at the use of chemicals, industrial practices, and pollution and how they affect the environment and human health, and raised questions about the long-standing belief that scientific progress is, without exception, good for humanity. Although *Silent Spring* was challenged by the pesticide industry, the public reaction to Carson's book eventually led to the establishment of a presidential commission charged with studying the effects of pesticides; the use of DDT was banned in the United States in 1972.

In 1989 Bill McKibben, a prolific nature writer, environmental commentator, and historian, also altered public perceptions with his book *The End of Nature*. By addressing such environmental issues as pollutants, acid rain, the greenhouse effect, and depletion of the ozone layer, McKibben raised awareness of the impacts of human activities on the earth's atmosphere and climate.

C. J. Walsh

FURTHER READING

Buell, Lawrence. *The Future of Environmental Criticism: Environmental Crisis and Literary Imagination*. Malden, Mass.: Blackwell, 2005.

Finch, Robert, and John Elder, eds. *Nature Writing: The Tradition in English*. New York: W. W. Norton, 2002.

Lyon, Thomas J. *This Incomparable Land: A Guide to American Nature Writing*. Rev. ed. Minneapolis: Milkweed Editions, 2001.

McKibben, Bill, ed. *American Earth: Environmental Writing Since Thoreau*. Des Moines, Iowa: Library of America, 2008.

Operation Backfire

CATEGORIES: Activism and advocacy; animals and endangered species

IDENTIFICATION: Criminal investigation led by the Federal Bureau of Investigation into a series of destructive actions carried out in the western United States in the 1990's in the name of environmental and animal rights causes

DATES: 2004-2006

SIGNIFICANCE: Operation Backfire is the largest and most highly publicized federal investigation of "ecoterrorism" thus far undertaken in the United States. The operation and subsequent trials fueled national debate over the definition of "terrorism" and the appropriate use of domestic surveillance.

Some critics regard Operation Backfire as an attempt to disrupt environmental and animal rights activism.

In 2004 the Portland, Oregon, field office of the Federal Bureau of Investigation (FBI) consolidated seven independent yet related investigations into a single major case that it code-named Operation Backfire. Counting the precursor investigations, Operation Backfire lasted nine years and was assisted by local, state, and federal law-enforcement agencies, including the Bureau of Alcohol, Tobacco, and Firearms. As a result of Operation Backfire's findings, eleven people were initially indicted; by 2010 that number had risen to seventeen.

Those indicted were purported to belong to a Eugene, Oregon-based cell of the Animal Liberation Front (ALF) and the Earth Liberation Front (ELF)—decentralized "direct action" animal rights and environmentalist groups considered to be terrorist organizations by the FBI—known as the Family. The charges included arson, conspiracy, use of destructive devices, and destruction of an energy facility. The crimes, which occurred between 1996 and 2001 in California, Colorado, Oregon, Washington, and Wyoming, resulted in approximately $48 million in property damages. Notably included was a highly publicized arson attack on a ski resort in Vail, Colorado, that was carried out to protest a planned expansion that would encroach on habitat of the endangered lynx and resulted in $12 million in damages.

While the indicted parties initially claimed innocence, fifteen of seventeen eventually pled guilty in federal courts. They were sentenced to prison in 2007 for periods ranging from 37 to 188 months; U.S. District Court Judge Ann Aiken imposed extended sentences owing to "terrorism enhancement" penalties. The alleged "mastermind" of the Family, Bill "Avalon" Rodgers, committed suicide in a Flagstaff, Arizona, jail in late 2005 before he could be transferred to Oregon.

Operation Backfire is not without its critics. While John Lewis, a top FBI official, declared in 2005 that "the No. 1 domestic terrorism threat is the ecoterrorism, animal-rights movement," some question the merit and the motive of this designation. In a May, 2005, congressional committee hearing, Senator Frank Lautenberg of New Jersey argued that Americans should not allow themselves to be blinded to more serious terrorist threats that take innocent lives by focusing on the illegal actions of ALF and ELF, be-

Types of Ecoterrorism

Category of Ecoterrorism	Percent of Incidents, 1993-2004
Vandalism	77.0
Arson	12.6
Assault and bodily harm	2.0
Bombings	1.1
Other	7.3

cause "not a single incident of so-called environmental terrorism has killed anyone." Senator Jim Jeffords of Vermont expressed similar skepticism. In a press release on Operation Backfire, the National Lawyers Guild said that the imposition of "draconian sentences" for property damage offenses without intent to harm is an unconstitutional and disproportionate punishment. Some have argued that Operation Backfire is part of a broader "Green Scare" (alluding to the twentieth century anticommunist Red Scare), which they allege is an attempt to suppress animal rights and environmental activism by exploiting public fears of terrorism, with the aim of maintaining the political and corporate status quo.

Defending Operation Backfire in 2006, FBI director Robert Mueller stated that "terrorism is terrorism—no matter what the motive." Mueller added that Operation Backfire dealt "a substantial blow" to domestic ecoterrorism and should have a "dramatic impact on persons who contemplate these crimes."

Joel P. MacClellan

Further Reading

Bishop, Bill. "Operation Backfire." *Register Guard*, April 15, 2006, A1.

Potter, Will. "The Green Scare." *Vermont Law Review* 33, no. 679 (2009): 671-687.

Scarce, Rik. *Eco-Warriors: Understanding the Radical Environmental Movement.* Updated ed. Walnut Creek, Calif.: Left Coast Press, 2006.

Osborn, Henry Fairfield, Jr.

CATEGORIES: Activism and advocacy; animals and endangered species; population issues

IDENTIFICATION: American naturalist and conservationist

BORN: January 15, 1887; Princeton, New Jersey

DIED: September 16, 1969; New York, New York

SIGNIFICANCE: Through his work with the New York Zoological Society and his writings, Osborn promoted the preservation of endangered species and their habitats and also raised public awareness of the dangers of human overpopulation.

Henry Fairfield Osborn, Jr., was one of five children of Henry Fairfield Osborn and Lucretia Thatcher Perry. The senior Osborn was a professor of paleontology at Princeton University at the time of Osborn's birth and eventually became the president of the American Museum of Natural History and founder of the New York Zoological Society. The younger Osborn graduated from the Groton School in 1905 and from Princeton University with a bachelor of arts in 1909. He then studied for one year at Cambridge University in England.

After finishing his year in Cambridge, Osborn traveled and held several jobs, including laying track for the railroad in Utah. During World War I he served as a captain in the 351st Field Artillery of the American Expeditionary Force. After returning from the war, Osborn worked in the investment banking business on Wall Street. He married Marjorie M. Lamond on September 8, 1914, and together they had three daughters. In 1922, as he continued working in investments, Osborn became a trustee of the New York Zoological Society. The paleontology field trips that he had taken with his father from an early age had convinced him that his real vocation lay in natural science. Osborn left the banking field in 1935 to become the secretary of the New York Zoological Society. He assumed the presidency of the society in 1940 and stayed in that position until 1968, when his health failed. Under his direction the Bronx Zoological Park was greatly improved, and the Marine Aquarium at Coney Island was created. In connection with his fund-raising for these causes, he has been called "the greatest showman since Barnum."

Osborn used his position in the Zoological Society to become a strong advocate of the preservation of endangered species and their habitats. His was also an early voice warning of the dangers of human population growth. He presented his ideas on this topic forcefully in his two books, *Our Plundered Planet* (1948) and *The Limits of the Earth* (1953). He also pushed his ideas about overpopulation as a member of the Conservation Advisory Committee of the U.S.

> ### Plundering the "Good Earth"
>
> *Henry Fairfield Osborn, Jr., opens* Our Plundered Planet *(1948) with the following cautionary words:*
>
> There is beauty in the sounds of the words "good earth." They suggest a picture of the elements and forces of nature working in harmony. The imagination of men through all ages has been fired by the concept of an "earth-symphony." Today we know the concept of poets and philosophers in earlier times is a reality. Nature may be a thing of beauty and is indeed a symphony, but above and below and within its own immutable essences, its distances, its apparent quietness and changelessness it is an active, purposeful, coordinated machine. Each part is dependent upon another, all are related to the movement of the whole. Forests, grasslands, soils, water, animal life—without one of these the earth will die—will become dead as the moon. This is provable beyond questioning. Parts of the earth, once living and productive, have thus died at the hand of man. Others are now dying. If we cause more to die, nature will compensate for this in her own way, inexorably, as already she has begun to do.

Department of the Interior and the Planning Committee of the Economic and Social Council of the United Nations. Osborn was often invited to speak at both public and scientific meetings, and he regularly attended national and international conservation conferences. In 1947 Osborn founded the Conservation Foundation; this organization was incorporated into the World Wildlife Fund in 1990. He was presented with the American Design Award in 1948 for his campaign to stop humanity from fighting a losing battle against nature.

Osborn suffered a slight stroke at seventy-nine years of age, which left him with a minor speech impediment. He continued to lecture and work, however, until his death in 1969 at the age of eighty-two.

Kenneth H. Brown

Further Reading

Collier, Paul. *The Plundered Planet: Why We Must—and How We Can—Manage Nature for Global Prosperity.* New York: Oxford University Press, 2010.

Shabecoff, Philip. *A Fierce Green Fire: The American Environmental Movement.* Rev. ed. Washington, D.C.: Island Press, 2003.

People for the Ethical Treatment of Animals

CATEGORIES: Organizations and agencies; activism and advocacy; animals and endangered species
IDENTIFICATION: Organization devoted to establishing and defending the rights of animals
DATE: Established in 1980
SIGNIFICANCE: People for the Ethical Treatment of Animals has grown to be one of the most powerful and effective animal rights organizations in the world, despite the fact that the organization's activities are often the subject of controversy.

Alex Pacheco, who founded People for the Ethical Treatment of Animals (PETA) with Ingrid Ward Newkirk in 1980, received initial notoriety for exposing cruelty to monkeys in a Silver Spring, Maryland, research laboratory. Although PETA began its campaign with only two people during the 1980's, by 2010 it had a worldwide membership of some two million people and annual expenditures of more than thirty million dollars. PETA's brand of radicalism and sophisticated use of the media are often the topics of national debate. During the 1980's PETA's major protest activities were centered on vivisection, factory farming, hunting, fishing, zoos, and circuses. Over time, the organization's concerns widened to include the fur industry and the use of animals in product testing, agricultural production, and biomedical research. PETA's campaign against the fur industry convinced designers such as Giorgio Armani, Ralph Lauren, and Calvin Klein, as well as supermodels such as Cindy Crawford and Christy Turlington, not to design or model clothing made of or incorporating fur.

The general principle upon which PETA was founded is simple but potent: Animals are not on earth for humans to eat, wear, or use for entertainment or experimentation. PETA asserts that advances in technology have enabled humans to make incredibly diverse substitutes for animals in all such uses. In advocating the end of all animal abuse, PETA maintains that violating animal rights is similar to violating human rights. In fact, the former is worse because animals cannot speak for themselves.

One tactic that PETA uses in attempting to stop individuals, organizations, or companies that violate animal rights is the undercover investigation. Many of PETA's investigations have revealed patterns of cruelty that have appalled the public and resulted in widespread public support for change as well as generous contributions of money to the organization. One of PETA's major goals is public education, which it accomplishes through the use of graphic visual images, expert testimony, media events, seminars, workshops, and lectures. PETA also conducts grassroots activities at colleges and universities.

PETA has been criticized for ecoterrorism and monkeywrenching, examples of which include illegal entry, vandalism, and theft of laboratory animals and equipment. Critics contend that PETA's philosophical stance is based on sentimentalism and narrow-minded dogmatism and that many of the organization's practices are akin to terrorism. They charge that PETA manipulates young people's emotions through misinformation disguised as education. Despite active opposition from industry and research communities, however, PETA has proven to be one of the most effective animal rights groups in the world. It has gained mainstream support from the public and from many powerful people involved in government, the entertainment industry, and humanitarian organizations.

Chogollah Maroufi

Further Reading

Best, Steven, and Anthony J. Nocella II, eds. *Terrorists or Freedom Fighters? Reflections on the Liberation of Animals.* New York: Lantern Books, 2004.

People for the Ethical Treatment of Animals. *All Animals Are Equal: The PETA Guide to Animal Liberation.* Washington, D.C.: Author, 2005.

Singer, Peter. *Animal Liberation.* 1975. Reprint. New York: HarperPerennial, 2009.

Pinchot, Gifford

CATEGORIES: Activism and advocacy; preservation and wilderness issues
IDENTIFICATION: American conservationist and forester
BORN: August 11, 1865; Simsbury, Connecticut
DIED: October 4, 1946; New York, New York
SIGNIFICANCE: As the first head of the U.S. Forest Service, Pinchot influenced national policy making concerning the conservation of natural resources as well as their management for human use.

After graduating from Yale University in 1889, Gifford Pinchot, whose family's fortune came in part from the lumber business, pursued further studies in forestry. Because no forestry school then existed in the United States, he trained at the French National Forestry School in Nancy in northeastern France for a year. His first forestry projects were for George W. Vanderbilt at the Biltmore estate and W. Seward Webb, Vanderbilt's brother-in-law, in the Adirondack Mountains in New York.

Pinchot believed that all forest resources should be made available for wise use by humans. He was critical of both Yellowstone National Park and the Adirondack State Forest Preserve because they did not permit logging. In his posthumously published autobiography, *Breaking New Ground* (1947), Pinchot states that "conservation is the foresighted utilization, preservation, and/or renewal of forests, waters, lands, and minerals, for the greatest good of the greatest number for the longest time." Conservationists led by Pinchot

Conservationist and forester Gifford Pinchot. (Library of Congress)

believed in the wise use of all the nation's resources. They were not preservationists who valued nature for nature's sake. Pinchot's chief ambition was to practice forestry on government land.

In 1891 the U.S. Congress passed a law allowing land to be removed from the public domain and set aside in forest reserves. At that time, the Bureau of Forestry was in the Department of Agriculture, but the forest reserves were administered through the General Land Office in the Department of the Interior. The reserves were run by political hacks holding patronage appointments. In 1897 the secretary of the interior asked Pinchot to travel west and report on the conditions of the reserves. Pinchot's assignment caused the head of the Bureau of Forestry to resign, and Pinchot was appointed to replace him in 1898. To supply their son with trained foresters, Pinchot's parents endowed the Yale Forest School (now the Yale School of Forestry and Environmental Studies), which opened in 1900 as a two-year postgraduate program. Pinchot still had no authority over the forest reserves.

Pinchot had advised Theodore Roosevelt on forestry when Roosevelt was governor of New York. Roosevelt's message to Congress after President William McKinley's assassination in 1901 favored placing the forest reserves under the authority of the Department of Agriculture. The transfer occurred in 1905 when the Forest Service was organized. Pinchot boasted that the forest reserves would be run to benefit poor settlers rather than rich people, a position that ultimately led to Pinchot's downfall.

With Roosevelt's approval, Pinchot brought the concept of conservation to national attention at the National Governors' Conference in 1907, the meeting of the National Conservation Commission in 1908, and the North American Conservation Conference in 1909. Pinchot's views were not supported by Roosevelt's handpicked successor, President William Howard Taft, or by Taft's secretary of the interior, Richard Ballinger. The resulting feud caused Taft to fire Pinchot. Although Pinchot remained active in public service (he served two terms as governor of Pennsylvania), he never fully regained his former position of national prominence.

Gary E. Dolph

FURTHER READING

Miller, Char. *Gifford Pinchot and the Making of Modern Environmentalism.* Washington, D.C.: Island Press, 2001.

Pinchot, Gifford. *The Conservation Diaries of Gifford Pinchot.* Edited by Harold K. Steen. Durham, N.C.: Forest History Society, 2001.

Wellock, Thomas R. *Preserving the Nation: The Conservation and Environmental Movements, 1870-2000.* Wheeling, Ill.: Harlan Davidson, 2007.

Population-control movement

CATEGORIES: Population issues; activism and advocacy

IDENTIFICATION: International development strategy that seeks to impose limits on population growth to improve quality of life

SIGNIFICANCE: The population-control movement is driven by concerns about the negative impacts of overpopulation. The movement has advocated the worldwide spread of family-planning services, particularly the distribution of contraceptives. Cultural and religious resistance can complicate family planning efforts. Some nations have used coercive methods to impose limitations on family size.

Birth control strategies have been used to regulate the timing and spacing of births for millennia, but the idea that humans should limit family size in order to improve quality of life originated with Thomas Malthus, an English cleric and economist who published *An Essay on the Principle of Population, as It Affects the Future Improvement of Society* in 1798. Nineteenth century neo-Malthusians were convinced that overpopulation caused poverty and that contraceptives should be provided to the poor, a position opposed by the medical community of the United States. Physicians triumphed when the Comstock Act was passed in 1873, making the sending of contraceptives and associated information through the U.S. mails illegal.

Support for birth control came from a diverse constituency, including woman suffragists, moral reformers, and advocates of eugenics (the science of improving human hereditary qualities). Some feminists advocated sexual abstinence rather than the use of "unnatural" contraceptives, but most believed that women should have the right to choose when to have a child. Margaret Sanger, a radical feminist and socialist, wrote extensively about birth control, set up clinics where women could seek contraceptive devices and advice, and recruited physicians. Sanger also founded the American Birth Control League in 1921—later renamed the Birth Control Federation of America (BCFA)—and helped organize the first World Population Conference in Geneva, Switzerland, in 1927. At the conference, many eugenicists, including Henry F. Osborn and Frederick Osborn, called for government intervention in birth control and sterilization of the "unfit."

In 1922 British sociologist and educator Sir Alexander M. Carr-Saunders published a book titled *The Population Problem* that laid the basis for "transition theory," a description of how fertility and mortality rates change during modernization. The evidence from European history indicated that the pattern of high fertility and mortality that preceded nutritional and health improvements was immediately followed by lowered mortality coupled to high fertility and rapid growth. Eventually the increased costs of raising children in an industrial, market economy led to smaller families, but in the poorer regions of Africa, Asia, and South America, large families were still preferred, even after mortality declined. Discovering how to speed the process of fertility decline became an important goal of demographers, eugenicists, and many feminists, such as Sanger and her British counterpart, Marie Stopes.

Eugenicist influence on the population-control movement was evident when the BCFA targeted African American doctors for assistance in spreading the family-planning message in 1939, a policy that was labeled racist by prominent African Americans. This effort followed the founding of birth control clinics in economically depressed Puerto Rico in 1935 by the government's relief agency. However, the Nazi atrocities during World War II tempered the more extreme positions of eugenicists, and the impoverishment of many educated and middle-class families during the Great Depression also confirmed that poverty was not necessarily a function of bad genes. Birth control information was thus added as part of the services provided to the poor by the U.S. government following the New Deal, and making that information more acceptable to a wider married constituency resulted in the renaming of the BCFA, which became the Planned Parenthood Federation of America in 1942.

Generating Support

By the 1950's and 1960's a number of wealthy businesspeople in the United States were waging a battle on behalf of birth control because of what they perceived as the evils of overpopulation. The major contributors to these efforts were the Ford and Rockefeller foundations. In 1952 John D. Rockefeller III founded and became the first president of the Population Council, an institution that aids countries in developing their own population policies by provide scientific research and grants. Hugh Moore, founder of the company that made Dixie Cups, sent his pamphlet *The Population Bomb* to ten thousand prominent citizens to garner support during the late 1950's, a time when most U.S. families practiced family planning and new contraceptive technologies were being developed. The first birth control pill (Enovid) was marketed in 1965.

Bringing the U.S. government into the picture was the big challenge. President Dwight D. Eisenhower formed the Committee to Study the U.S. Military Assistance Program in 1958 because of complaints that U.S. funds for overseas assistance put too much emphasis on military support at the expense of economic aid. Major General William H. Draper, Jr., the former director of the European Recovery Program, chaired the committee and was encouraged by Moore to study the population problem. Ansley Coale and Edgar Hoover's *Population Growth and Economic Development in Low-Income Countries* (1958) had just been published; this work showed that having too many children overtaxed family resources, reducing private investment. That economic development might be impeded by rapid population growth provided Draper with the rationale for using international assistance for fertility limitation, but his report was later disavowed by President Eisenhower and the Roman Catholic bishops of the United States. Draper later tried to convince President John F. Kennedy to take action, but Kennedy suggested that the private sector, particularly the Ford Foundation, fund population control.

During the 1950's a series of surveys (known as knowledge, attitudes, and practices, or KAP, surveys) measured the "unmet need" for contraception and found that most women would have smaller families if they could. Dramatic publicity about overpopulation during the early 1960's alarmed U.S. voters. Thus, by 1965, during the administration of President Lyndon B. Johnson, the federal government began a tradition of funding population programs through the U.S. Agency for International Development (USAID) and later also through the United Nations.

United Nations

In 1962 the United Nations invited member states to formulate population policies, and in 1966 it reemphasized the connection between population and socioeconomic factors. The United Nations also reinforced individual governments' rights in setting policy and the right of families to determine number, timing, and spacing of births. However, Western and Asian countries, particularly India, Sweden, and the United States, pressured the United Nations to take on a leadership role in instituting family-planning programs.

In 1967 United Nations secretary-general U Thant established the U.N. Trust Fund for Population Activities (renamed the U.N. Fund for Population Activities, or UNFPA, in 1987) with the goals of helping developing nations with population-related matters, expanding the UN's role in family planning, and pursuing new programs. He was assisted in his efforts by Reimert T. Ravenholt, head of USAID, who wanted contraceptives distributed worldwide in order to hasten economic development. By 1971 the trust fund was a recognized hub of the United Nations system, which also included support for population activities by the International Labour Organization, the Food and Agriculture Organization, the U.N. Educational, Scientific, and Cultural Organization (UNESCO), and the World Health Organization.

The first world conferences to study population issues were held in Rome, Italy, in 1954 and Belgrade, Yugoslavia, in 1965. However, the Third World Population Conference held in Bucharest, Romania, in 1974 was the first official government conference with an emphasis on policy rather than research. Delegates agreed that population issues needed to be addressed, but representatives from the Northern Hemisphere and those from the Southern Hemisphere were strongly divided over the relative importance of population planning versus economic development. Delegates from Southern countries believed that an inequitable distribution of resources and the heavy consumption patterns of wealthy Northern nations contributed as much to environmental deterioration as population growth by the poor.

Reproductive Politics

The Federal Office of Population Affairs was founded in 1970 and continues to provide federal money for family-planning services (except abortions) for poor Americans. In 1972 the Commission on Population Growth and the American Future advocated population planning as important for world stability. The commission believed that efforts should include access to abortion and limits to illegal immigration. President Richard Nixon would not approve the document, and debate intensified when the U.S. Supreme Court prohibited interference with a woman's right to an abortion during the first three months of pregnancy in *Roe v. Wade* (1973). At the 1984 Population Conference in Mexico City, the U.S. delegation reversed its support for family-planning measures because of concerns about coercive population-control measures, such as forced abortion and involuntary sterilization in some countries, notably the People's Republic of China. However, the World Plan of Action written at Bucharest was revised by delegates representing nations in general agreement that family-planning programs are useful whether economic development takes place or not.

Out of the 1984 conference came the Mexico City Policy, or Global Gag Rule, a U.S. policy implemented under President Ronald Reagan that denies federal funding to nongovernmental organizations that provide or promote abortions overseas. Under this policy, the United States no longer funded the International Planned Parenthood Federation or the UNFPA, a position that was not reversed until President Bill Clinton took office in 1993. One of President George W. Bush's first actions after his inauguration in 2001, however, was to reinstate the policy. Yet another reversal came almost exactly eight years later under President Barack Obama, who rescinded the policy a few days after being sworn in.

Abortion is a particularly contentious topic in the United States. While forced abortion and other coercive, abusive population-control measures overseas have been cited as the reason for the United States to withhold funding from international family-planning organizations, U.S. antiabortion forces tend to oppose the voluntary termination of pregnancies as well. The antiabortion movement that arose in the United States in the wake of the 1973 *Roe v. Wade* decision was initially led by the U.S. Catholic Church, but from the 1980's onward the movement increasingly became associated with Christian fundamentalism. The belief that life begins at conception and that abortion is therefore synonymous with murder motivates the most passionate opposition to the termination of pregnancies. Some even regard contraception as an immoral choice that defies the will of God. For these reasons, U.S. funding of international family-planning interests is a politically charged subject. Abortion and contraception tend to be similarly controversial in countries that are predominantly Roman Catholic or Muslim.

ONGOING DEBATES

No consensus exists on the interrelationships among population growth, environmental degradation, and economic development. Some feel population growth is a cause of poverty, while others would reverse that relationship. Women with the lowest education and wages and fewest opportunities outside the household tend to have the largest families. The perspective that gender inequality lies at the heart of the problem in the regions of the world that continue to exhibit high fertility has gained momentum, particularly evident at the International Conference on Population and Development in Cairo, Egypt, in 1994.

Many Southern Hemisphere nations have subsistence economies requiring labor-intensive strategies, and larger families are a rational choice where savings are difficult to put aside and children are productive assets who also provide security during old age. Children in rural India as young as six years old can look after domestic animals and younger siblings, and help with other tasks. In extended-family households, the costs of raising children are also shared because access to common lands expands as the household becomes larger. However, increasing urbanization and market pressures are forcing changes in traditional ownership of land so that these shared lands are decreasing in spite of ever-larger demands that put pressure on local environments.

Few doubt the importance of family-planning measures, and most Southern nations support and feel responsible for providing services to individual couples. However, reformers have tried to shift attention to women's health and empowerment because the population-control approach is believed to have resulted in ethical violations and coercive abuses. Furthermore, it is not always successful where children are an important source of labor. In the twenty-first century, population control efforts have placed emphasis on women's reproductive health, with increased sensitivity to local context. In particular, effecting societal change within communities in the developing world that practice child marriage has become a priority. Young women who receive an education, enjoy some financial independence, and marry later in life tend to be healthier, better mothers, give birth to healthier children, live less isolated lives, and be less likely to become trapped in physically abusive marriages.

Joan C. Stevenson
Updated by Karen N. Kähler

FURTHER READING

Brown, Lester R., Gary Gardner, and Brian Halweil. *Beyond Malthus: Nineteen Dimensions of the Population Challenge.* Sterling, Va.: Earthscan, 2000.

Eager, Paige Whaley. *Global Population Policy: From Population Control to Reproductive Rights.* Burlington, Vt.: Ashgate, 2004.

Harkavy, Oscar. *Curbing Population Growth: An Insider's Perspective on the Population Movement.* New York: Plenum Press, 1995.

Hartmann, Betsy. *Reproductive Rights and Wrongs: The Global Politics of Population Control.* Boston: South End Press, 1995.

Huggins, Laura E., and Hanna Skandera, eds. *Population Puzzle: Boom or Bust?* Stanford, Calif.: Hoover Institution Press, 2004.

Tobin, Kathleen A. *Politics and Population Control: A Documentary History.* Westport, Conn.: Greenwood Press, 2004.

Powell, John Wesley

CATEGORY: Land and land use
IDENTIFICATION: American geologist and explorer
BORN: March 24, 1834; Mount Morris, New York
DIED: September 23, 1902; Haven, Maine
SIGNIFICANCE: Powell contributed significantly to scientific knowledge of the American West in the mid-nineteenth century, and his ideas regarding environmental policy are recognized as being ahead of their time.

Best known for his explorations of the western United States, John Wesley Powell contributed to both the romantic and the scientific views of those lands. Many stories have been told of his exploits in leading surveying expeditions through uncharted territory, especially his trip down the Grand Canyon in 1869. Newspapers across the United States reported on the progress of Powell's travels as he led that first exploration of the Grand Canyon by Euro-

Geologist and explorer John Wesley Powell. (NPS)

pean Americans. In the twenty-first century, people preparing for recreational white-water rafting trips still read his accounts of the challenges he faced on the Colorado and Green rivers. The personal drama of the stories was heightened by the fact that Powell had lost his right arm in the Battle of Shiloh during the Civil War.

Powell's first contributions to scientific knowledge of the American West were the materials he returned to the Smithsonian Institution in Washington, D.C., and the reports he made to the U.S. Congress. Among the most enduring of his environmental science contributions was his initiative in generating topographic maps of the United States. He envisioned a set of maps that would cover the entire country, but the project was too big for his lifetime. Still, that initiative continues to have direct impacts on people who enter public lands, whether for recreational or commercial purposes.

Wallace Stegner's 1954 biography *Beyond the Hundredth Meridian: John Wesley Powell and the Second Opening of the West* brought Powell's views back to the attention of environmental management officials. Stegner presented Powell not only as a leading figure in the organization of science in government but also as an environmental policy visionary who was ahead of his time. Powell's *Report on the Lands of the Arid Region of the United States* (1878), written a few years before he became director of the U.S. Geological Survey in 1881, advised against the use of the checkerboard grid—the standard approach to land surveying then being used in the eastern United States—in the surveying of the West. Arguing that the lack of water in the West needed to be the dominant factor in dividing land for agricultural purposes, Powell promoted an approach in which 65 hectares (160 acres) would not be standard. He saw 32.4 hectares (80 acres) as a sufficient amount of irrigable land for a homestead and thought that 1,036 hectares (2,560 acres) might be needed for pasturage. Though the National Academy of Sciences supported his recommendations, they were not accepted by Congress.

The value of Powell's influence on federal land policy remains in dispute, but in the late twentieth century the Bureau of Reclamation called him "the father of irrigation development," and at least two secretaries of the U.S. Department of the Interior (Stewart Udall and Bruce Babbitt) have claimed that they were inspired by Powell.

Larry S. Luton

FURTHER READING
Aton, James M. *John Wesley Powell: His Life and Legacy.* Salt Lake City: Bonneville Books, 2010.
Worster, Donald. *A River Running West: The Life of John Wesley Powell.* New York: Oxford University Press, 2001.

Public opinion and the environment

CATEGORIES: Philosophy and ethics; activism and advocacy
SIGNIFICANCE: Public opinion is one of the primary factors determining the degree of environmental protection in a society.

A clean environment is a highly desirable amenity, and on occasion very important in terms of long-term health. It can also be expensive in many ways. As a result, except in extreme cases, environmental protection is a luxury rather than an immediate necessity. Everyone wants a clean local environment, as long as the price they pay personally is reasonable. This makes spectacular events, such as the Cuyahoga River fires of 1952 and 1969 and large oil spills, especially significant.

HISTORICAL EXAMPLES

It has long been noticed that more affluent populations have cleaner environments. A 1992 World Bank study reported that concentrations of airborne particulates started to decline with a per-capita gross national product (based on purchasing power parity) of $3,280, sulfur dioxide at $3,670, and fecal coliform bacteria in river water at $1,375. Access to safe water and adequate sanitation is thus more immediately important than access to clean air (which is less visible). Similar data reveal that U.S. sulfur dioxide emissions per capita peaked in 1920, particulates and carbon monoxide have declined steadily since 1945, and nitrogen oxides peaked in 1980. A 1992 study by the World Health Organization found that air pollutants were generally lower in most megacities (except for Los Angeles) in the noncommunist developed world than in developing nations.

This situation is caused in part by the fact that more affluent societies can afford expensive water treatment, plumbing, and air-pollution controls. For example, from 1972 to 1994 annual pollution-control expenditures in the United States (in 1986 dollars) increased steadily from $26 to $127 billion. This grew especially rapidly starting in 1970 (when the Environmental Protection Agency was created), the year after the Santa Barbara, California, oil spill and the last Cuyahoga River fire. Part of the difference between affluent and developing societies (especially noticeable in pollutant levels), however, results from the fact that less affluent societies have more urgent concerns.

After the fall of the Communist dictatorship of the Soviet Union, Green parties did well in many places in the former Soviet countries, especially in local elections. Environmental quality there was often atrocious, demonstrated by such notorious problems as the massive shrinking of the Aral Sea and catastrophic levels of air and water pollution from heavy industry that led to high levels of infant mortality and birth defects, food contamination, and environmentally induced illnesses. In addition, attempts to fix old Soviet industrial facilities often led to their closure, at least temporarily, which resulted in serious loss of production even of necessary supplies (such as pharmaceuticals) and led to a serious backlash against environmental protection (much of it by the authorities, but ultimately also by voters).

TWENTY-FIRST CENTURY ISSUES

The struggle to save endangered species reveals the important of public opinion. An estimated 95 percent of endangered species are tiny, obscure plants (including algae and fungi) or invertebrates (insects, worms, and mollusks), mostly tropical. Groups that seek to protect endangered species, however, generally use more popular animals, especially birds (such as the spotted owl and the whooping crane) and large mammals (such as polar bears and various whale species), as emblems.

Another example of the importance of public opinion can be seen in the effects of the severe economic downturn that started in 2008. During this crisis, environmental activism lost much of its popularity with the American public. A spill as spectacular as the one created by the blowout of the BP *Deepwater Horizon* oil rig in 2010 may in previous decades have led to strong popular demand to end offshore oil drilling (as happened after the 1969 spill off the shores of Santa Barbara). Instead, polls showed strong popular support for continued drilling, not only from those involved in the drilling but also often from those whose occupations were harmed by the spill.

The most significant environmental dispute in the United States involves global warming, with debates concerning whether the cause is primarily anthropogenic (human-created) or natural and whether the results will be catastrophic. Both sides rely heavily on emotional appeals rather than on rational discussion of the science. Skeptics primarily discuss the expense of measures to reduce the greenhouse gas emissions linked with global warming, a tactic that is especially effective among those (such as coal miners) most affected by proposed preventive measures. Some supporters of taking action to reduce greenhouse gases greatly exaggerate the likely effects of global warming and ignore (and even smear as corrupt) any scientific dissent; some have been accused of falsifying data.

CAUTIONARY NOTES

Acid rain provides a very instructive lesson regarding the influence of public opinion on environmental policy. Starting around 1980, acid rain was a major theoretical concern. The fear was that acid rain was causing significant damage to forests, eroding buildings, and causing increasingly acidic lakes (often leading to the disappearance of fish). In response to public concerns, the United States set up in the National Acidic Precipitation Assessment Project, which after ten years of study concluded that acid rain is primarily an aesthetic concern. Acidic lakes result primarily from acidic deposits from land; many such lakes were found to have been only temporarily neutral or alkaline as a result of heavy lumbering, which had since ceased. Acid rain causes only minor damage to high-altitude forests, while actually fertilizing the soil with additional nitrogen and sulfur. (Later, many global warming scientists concluded that sulfate haze had also helped cool the atmosphere, explaining why global temperatures had increased less than expected.)

During the 1980's, however, the news media operated in standard crisis mode, making the unproven theory of serious damage caused by acid rain appear to be a major proven crisis. As a result, significant pollution controls were put into place to manage a relatively minor problem. It was certainly desirable for these controls to be instituted, but the resources available for pollution control are not unlimited, and it is possible that those resources could have been better used to deal with other, more serious, environmental problems.

Timothy Lane

FURTHER READING

Bailey, Ronald, ed. *The True State of the Planet.* New York: Free Press, 1995.

Feshbach, Murray, and Alfred Friendly, Jr. *Ecocide in the USSR: Health and Nature Under Siege.* New York: Basic Books, 1992.

Lomborg, Bjørn. *The Skeptical Environmentalist: Measuring the Real State of the World.* New York: Cambridge University Press, 2001.

Mooney, Chris. *Storm World: Hurricanes, Politics, and the Battle over Global Warming.* Orlando, Fla.: Harvest Books, 2008.

Murray, Iain. *The Really Inconvenient Truths.* Washington, D.C.: Regnery, 2008.

Rainforest Action Network

CATEGORIES: Organizations and agencies; activism and advocacy; forests and plants
IDENTIFICATION: Nonprofit organization that works to protect tropical rain forests and other endangered forests and the rights of the peoples native to those forests
DATE: Founded in 1985
SIGNIFICANCE: Through its programs and projects, in particular consumer boycotts and letter-writing campaigns, the Rainforest Action Network raises awareness of the environmental impacts of deforestation and puts pressure on governments and corporations to end practices that endanger the world's forests.

The Rainforest Action Network (RAN) publicizes the environmental dangers associated the destruction of rain forests and old-growth forests, focusing public attention on the actions of companies involved in the degradation of the world's forests. RAN achieves its conservation mission through education and direct grassroots activities, with support from activists working in countries with rain forests. RAN organizes product boycotts to influence corporate executives and uses letter-writing campaigns, petition drives, and nonviolent demonstrations to influence public policy makers. RAN also develops coalitions among scientific, environmental, and grassroots organizations worldwide; holds conferences and seminars; and provides technical and financial assistance to native communities and nongovernmental organizations (NGOs) in rain-forest countries.

At the international level, RAN members participate in letter-writing campaigns targeting the leaders of countries that permit the destruction of rain forests. Information sharing and coordination of activities is facilitated through RAN's cooperative alliances with other environmental and human rights groups in more than sixty countries. At the local level, RAN members are organized within grassroots organizations known as Rainforest Action Groups. Members of these local organizations are encouraged to write letters to policy makers, coordinate nonviolent demonstrations, organize product boycotts, and participate in educational and direct-action campaigns. Through its Protect-an-Acre program, RAN supports local organizations within rain-forest countries that initiate projects to protect the ecological or cultural integrity of forest communities.

In its first notable success, RAN led a boycott of Burger King fast-food restaurants across the United States to raise public awareness concerning Burger King's purchase of beef from companies involved in expanding pastureland for cattle at the expense of rain forests. This campaign, which ran from 1985 to 1987, led to a 12 percent drop in Burger King's sales and prompted company officials to cancel $35 million in contracts for beef raised in Central America and discontinue the company's purchase of beef fed on former rain-forest lands. In 1998, RAN's success in leading a boycott of products produced by the Mitsubishi Corporation encouraged corporate executives to discontinue company practices that were harmful to rain forests and their native cultures.

The public pressure resulting from other RAN campaigns has also influenced many large companies to change their policies concerning the purchase and resale of old-growth wood products from U.S. forests. For example, in 1999 Home Depot stores made a commitment to stop selling old-growth redwood. Other companies from which RAN has won concessions include Scott Paper, Boise Cascade (a manufacturer of paper and other wood products), Occidental Petroleum, and Goldman Sachs investment bank. RAN has also encouraged organizations such as the World Bank to deny funding for companies involved in ecologically destructive activities within rain forests. Because timber harvesting is the leading agent in rain-forest destruction, RAN has become a strong advocate for the use of sustainable alternatives to pulp, paper, and tropical wood in furniture and building construction.

Thomas A. Wikle

FURTHER READING

Gunther, Marc. "The Mosquito in the Tent: A Pesky Environmental Group Called the Rainforest Action Network Is Getting Under the Skin of Corporate America." *Fortune*, May 31, 2004, 158-162.

Holzer, Boris. "Transnational Protest and the Corporate Planet: The Case of Mitsubishi Corporation Versus the Rainforest Action Network." In *Environmental Sociology: From Analysis to Action*, edited by Leslie King and Deborah McCarthy. Lanham, Md.: Rowman & Littlefield, 2005.

Place, Susan E., ed. *Tropical Rainforests: Latin American Nature and Society in Transition*. Rev. ed. Wilmington, Del.: Scholarly Resources, 2001.

Sagebrush Rebellion

CATEGORY: Land and land use
DEFINITION: A U.S. movement that called for state, rather than federal, management of public lands
SIGNIFICANCE: The Sagebrush Rebellion evidenced the growing conflict between environmentalism and the economic exploitation of public lands. Although the movement did not survive past the early 1980's, it succeeded in changing the policy debate regarding public lands.

The Sagebrush Rebellion, which originated in the western United States during the late 1970's and early 1980's, was part of a larger historical trend that began during the nineteenth century as settlers in the West struggled to gain control over public lands. Movements opposing federal management of these lands had arisen periodically, as they would again during the 1990's in the form of the wise-use movement. The Sagebrush Rebellion received national attention during the 1980's in part because of President Ronald Reagan's support for the movement and because it was part of a larger movement that demanded the reduction of the federal government's involvement in American life. The Sagebrush Rebellion represented a response to the environmental movement that gained ground during the 1970's.

The federal government owns nearly 50 percent of the land in the western United States. In response to growing environmental activism, government regulations concerning the use of these public lands increased significantly during the 1970's. Several of these regulations threatened lumbering, cattle grazing, and other commercial activities that were permitted on public lands. Many westerners regarded the regulations as unfair, claiming that they stifled development and slowed economic growth in the region. Some maintained that the federal government was limiting opportunities in the West in order to protect established industries in the eastern United States.

In response, these westerners advocated the transfer of land management from the federal government to state governments, which were more amenable to development. They formed organizations such as the League for the Advancement of States' Equal Rights to advance their cause. During the late 1970's two members of Congress from western states offered legislation at the federal level to transfer land management to the states; however, these initiatives gained little support.

The movement garnered national attention in 1979 when the Nevada legislature passed Assembly Bill 413, commonly known as the Sagebrush Rebellion Act. Nevada's legislators maintained that the huge expanse of federal lands within the state's borders (83 percent of the land in Nevada belonged to the federal government) put the state at a disadvantage in comparison with other states. They claimed the right to manage the 19.4 million hectares (48 million acres) of land then under the authority of the U.S. Bureau of Land Management (BLM), which constituted almost all the federal lands in the state. After passage of the act, Nevadans made no effort to seize control of the public lands. Instead, the state appropriated funds to initiate a lawsuit against the federal government.

The so-called Sagebrush Rebels soon counted several victories throughout the region. Four western states enacted similar measures, with the Arizona legislature overriding Governor Bruce Babbitt's veto. In 1982 Alaskans showed their support for the so-called Tundra Rebellion when more than 70 percent of voters supported efforts to take control of that state's BLM lands. Success was not guaranteed—the Sagebrush Rebels saw legislative or executive measures supporting their cause fail in seven western states—but the notion of state control of public lands was clearly popular with western voters.

The movement gained a powerful boost in 1980 when president-elect Reagan declared himself to be a Sagebrush Rebel. As president he followed through on his rhetoric, nominating another "rebel," James Watt, for the cabinet position of secretary of the interior. Emboldened by the changed political climate in Washington, D.C., in May, 1981, Senator Orrin Hatch of Utah again offered legislation requiring state management of BLM lands and lands managed by the U.S. Forest Service. As in the past, the proposal did not gain sufficient support in Congress. The failure of Hatch's legislation indicated that the Sagebrush Rebellion lacked support from key politicians. Although some government officials agreed with the westerners' complaints, others merely made campaign promises that they had no intention of fulfilling.

Weaknesses of the Movement

The Sagebrush Rebellion suffered from several weaknesses. First, the state initiatives offered a potent rallying point for the movement, but few politicians or other observers took them seriously. The courts clearly would not approve measures that intruded on the powers of the federal government. Land transfers had to take place at the federal level, and despite Reagan's stated enthusiasm for the rebellion, key members of his administration opposed the notion of state or federal government ownership of lands. They favored the sale of public lands to private concerns. In the early 1980's many Sagebrush Rebels allied themselves with environmentalists to thwart the privatization movement.

Observers also noted that the rebellion faced opposition from the very groups that it was intended to assist. In many western states ranchers complained about federal regulation but benefited from the low cost of grazing on public lands. Mining companies also feared that the transfer of public land to state control might raise their costs and threaten their mineral rights. Some western state governments estimated that the costs of managing the lands would exceed the income the lands would generate while under state control. Thus powerful western interests were arrayed against the Sagebrush Rebellion.

The movement was short-lived, lasting only from the late 1970's to 1983, when the debate over privatization effectively killed it. If it is assessed by its goals, the Sagebrush Rebellion failed because federal lands remained in the public domain. However, the movement did change the policy debate regarding public lands. It advanced the ethic of multiple-use management, which remained an important issue in public land policy from the 1980's onward. It also provided public support for the Reagan administration's efforts to scale back environmental regulations.

Thomas Clarkin

Further Reading

Cawley, R. McGreggor. *Federal Land, Western Anger: The Sagebrush Rebellion and Environmental Politics.* Lawrence: University Press of Kansas, 1993.

Davis, Charles, ed. *Western Public Lands and Environmental Politics.* 2d ed. Boulder, Colo.: Westview Press, 2001.

Echeverria, John D., and Raymond Booth Eby, eds. *Let the People Judge: Wise Use and the Private Property Rights Movement.* Washington, D.C.: Island Press, 1995.

Graf, William L. *Wilderness Preservation and the Sagebrush Rebellions.* Savage, Md.: Rowman & Littlefield, 1990.

Nelson, Robert H. *Public Lands and Private Rights: The Failure of Scientific Management.* Lanham, Md.: Rowman & Littlefield, 1995.

Platt, Rutherford H. *Land Use and Society: Geography, Law, and Public Policy.* Rev. ed. Washington, D.C.: Island Press, 2004.

Robbins, William G., and James C. Foster, eds. *Land in the American West: Private Claims and the Common Good.* Seattle: University of Washington Press, 2000.

Sale, Kirkpatrick

CATEGORY: Activism and advocacy
IDENTIFICATION: American journalist, historian, and environmental writer
BORN: June 27, 1937; Ithaca, New York
SIGNIFICANCE: In a long career as journalist and activist, Sale has helped shape the modern environmental movement through his writings emphasizing human scale, bioregionalism, decentralization, and a thoroughgoing critique of technology and the idea of progress.

Kirkpatrick Sale was educated at Quaker-associated Swarthmore College and at Cornell University, where he majored in history and journalism. This educational background underlies Sale's long-standing efforts to forge links between environmental and social justice concerns—efforts reflected in the many books he has produced in his career as journalist and historian, from *The Land and People of Ghana* (1963) and *SDS* (1973) to *Rebels Against the Future: The Luddites and Their War on the Industrial Revolution—Lessons for the Computer Age* (1995) and *After Eden: The Evolution of Human Domination* (2006). Although sometimes portrayed as a simple Luddite, anarchist, and secessionist, he has served as editor with *The New York Times Magazine* and *The Nation* and is a member of both the E. F. Schumacher Society and the American Association for the Advancement of Science.

The trajectory of Sale's career as environmental writer and activist can perhaps be best understood in terms of the importance of a Schumacher-influenced "appropriateness" in Sale's thought and writings—most important, appropriateness in tools, scales, and relations among humans and of humans with other species and the nonhuman environment generally. His works trend toward a critique of what might be called the technoglobal dominance of the human species, the dangerous failings of which are, at root, all ultimately failures of appropriateness.

Although for Sale technology is never neutral, tools in themselves are less an issue than the prior history built into them by the culture that made them. Cultures that increasingly elevate the technological and material—to the exclusion of environmental, social, civic, and other "irrelevant" values—produce increasingly inappropriate and unsustainable technologies. As a bioregionalist who believes that human beings live best when they are aware of and live within the constraints of the regions in which they find themselves, Sale views the increasingly global scale of human technological culture also as increasingly inappropriate to long-term sustainability. The increasingly dominant (and increasingly inappropriate) power position of a relatively few human beings within the human species and the increasingly dominant position of *Homo sapiens* as the "crown of creation" in the natural world are at odds with horizontal and decentralized power relations among human beings and between the human species and the rest of nature—these are key elements of Sale's thought concerning what is appropriate and sustainable.

The broad "revolutions" of the past seventy thousand years—the introduction of big-game hunting among hunter-gatherers, the agricultural revolution, the Industrial Revolution, the digital revolution—are, for Sale, all intensifications of the same pattern of increasing inappropriateness in human tools, scales, and relations, a vast "boom" leading inevitably to a catastrophic "bust" on the same scale—unless action is taken in time.

Whether he is discussing the industrial capitalism of the past few centuries or the growing technoglobal dominance of *Homo sapiens* over the past seventy thousand years, Sale presents a message that mingles apocalypse and optimism. The forces he presents as arrayed against the change to a more appropriate way of living always seem to be unstoppably and invincibly leading humankind to global catastrophe—but this is precisely where the relevance of history to the future comes into play. Although humanity may be suffering a second, digital revolution in the present, we still have before us the model of the nineteenth century Luddite revolt against the first industrial revolution. Although we live in a technoglobally dominant *Homo sapiens* world, we still have the nearly two-million-year run of the more appropriate world of *Homo erectus* to inspire us. Sale, as historian, reminds us that the past was not the same as the present, and this makes legitimate the hope for appropriate change, since the future need not be merely a continuation of the present.

Howard V. Hendrix

FURTHER READING

Katz, Eric, Andrew Light, and David Rothenberg, eds. *Beneath the Surface: Critical Essays in the Philosophy of Deep Ecology.* Cambridge, Mass.: MIT Press, 2000.

Sale, Kirkpatrick. *Dwellers in the Land: The Bioregional Vision.* Athens: University of Georgia Press, 2000.

Schumacher, E. F. *Small Is Beautiful: Economics As If People Mattered.* 1973. Reprint. Point Roberts, Wash.: Hartley & Marks, 1999.

Thayer, Robert L., Jr. *LifePlace: Bioregional Thought and Practice.* Berkeley: University of California Press, 2003.

SANE

CATEGORIES: Organizations and agencies; activism and advocacy; nuclear power and radiation

IDENTIFICATION: Organization of environmental activists, scientists, and pacifists formed with the aim of protesting nuclear weapons testing

DATE: Founded in 1957

SIGNIFICANCE: SANE experienced some success in increasing public awareness regarding the effects of nuclear testing on the environment.

In April, 1957, prominent antinuclear activists and pacifists, led by *Saturday Review* editor Norman Cousins, met in Philadelphia; the ad hoc organization formed at this meeting evolved into SANE. In September, the group's officials named their organization the National Committee for a Sane Nuclear Policy. The formation of what soon came to be known simply as SANE was announced in *The New York Times* on November 15, 1957, in an advertisement with the headline "We Are Facing a Danger Unlike Any Danger That Has Ever Existed." The advertisement, which was sponsored by such notables as theologian Paul Tillich, social critic Lewis Mumford, novelist James Jones, and humanitarian Eleanor Roosevelt, brought the group many new members and donations.

SANE was at first an informal venture meant to make the public aware of the dangers of nuclear weapons testing, but it proved so popular that it became a permanent organization. The group's initial membership consisted largely of scientists, writers, and other professionals, but as the antinuclear movement grew, thousands of other people joined. By mid-1958, SANE had approximately twenty-five thousand members and more than one hundred chapters.

Despite Cold War opposition, SANE and other groups pressed their campaign against nuclear testing. In August, 1958, the United States followed the lead of the Soviet Union and voluntarily suspended nuclear tests. The two countries further agreed to meet in Geneva, Switzerland, to begin negotiations for a test ban treaty.

In the early 1960's, SANE came under attack by members of the U.S. Congress for the organization's alleged harboring of communists. Investigations by SANE leaders established that some members did in fact have communist affiliations, and though congressional investigations eventually exonerated SANE's leadership of wrongdoing, the effects of the investigations on SANE were disastrous. Membership declined drastically, and several prominent members and sponsors resigned.

In 1961, when the Soviet Union resumed the testing of nuclear weapons, SANE condemned the action and called for international protests. The organization also urged the United States to refrain from following the Soviet example, but to no avail.

Such concerns became secondary during the Cuban Missile Crisis of October, 1962, when the world stood on the brink of nuclear war. In the aftermath, support for a treaty between the United States and the Soviet Union gained new life, and the Limited Test Ban Treaty was signed on August 5, 1963. With the signing of the treaty, SANE's initial goals were partially achieved.

SANE leaders subsequently chose to direct the group's energies to opposing the Vietnam War, but this approach proved divisive; in 1967, many key SANE officials, including Cousins and the executive director Donald Keys, resigned from the organization. SANE's membership and influence subsequently declined precipitately, and in 1969 the group changed its focus to campaign against the construction of antiballistic missile systems. The organization did not experience a resurgence until the early 1980's, when the nuclear freeze movement gathered momentum. In the late 1980's, SANE merged with another organization, the Nuclear Weapons Freeze Campaign, to become SANE/Freeze, the largest peace-promoting organization in American history. SANE/Freeze was renamed Peace Action in 1993.

Alexander Scott

FURTHER READING

Giugni, Marco. *Social Protest and Policy Change: Ecology, Antinuclear, and Peace Movements in Comparative Perspective.* Lanham, Md.: Rowman & Littlefield, 2004.

Wittner, Lawrence S. *Confronting the Bomb: A Short History of the World Nuclear Disarmament Movement.* Stanford, Calif.: Stanford University Press, 2009.

Save the Whales Campaign

CATEGORIES: Activism and advocacy; animals and endangered species
IDENTIFICATION: A sustained advocacy effort on behalf of the protection of whales
DATE: Initiated in 1971
SIGNIFICANCE: The efforts of the variety of individuals and organizations involved in the Save the Whales Campaign have been influential in bringing about a ban on commercial whaling and in arousing international condemnation of the activities of pirate whalers and whale meat smugglers.

The unchecked commercial hunting of whales over several centuries in both open oceans and coastal waters caused the depletion of whale populations worldwide. Once harvested for their oil, whales later became prized for their meat, considered to be a delicacy in Japan and a few other countries. Many whale species approached extinction because of rampant overhunting.

The International Whaling Commission (IWC) was founded in 1946 to prevent further exploitation of whale populations, but instead it presided over some of the worst excesses in whaling's history. In 1971 the Animal Welfare Institute (AWI) joined with several other organizations to launch the Save the Whales Campaign in an effort to end the harvesting of whales. Pursuing the theme of this campaign, the United Nations Conference on the Human Environment in 1972 called for a ten-year moratorium on commercial whaling. In 1974 the IWC decided to regulate whaling according to the principle of maximum sustainable yield. Whenever a species of whale dropped below the optimal population for such a yield, the IWC instituted a ban on hunting that species so that the population could recover. In 1974 the blue whale, bowhead whale, and right whale populations reached low levels and were immediately protected. Because of the difficulty of obtaining reliable data and enforcing catch limits, the Save the Whales Campaign encouraged the IWC in 1982 to adopt an indefinite moratorium on commercial whaling; such a moratorium took effect in 1986 when the IWC instituted its international whaling ban.

During the 1990's the philosophy of the Save the Whales Campaign expanded to include not only the conservation of the whale population but also the issues of animal rights and the aesthetic value of observing whales. Many people worldwide simply believe that it is wrong to kill and eat such large, unique animals, and the World Wide Fund for Nature and others have pointed out that whales must be preserved for observation because of their intrinsic value as mammals of great intelligence.

In opposition to the Save the Whales Campaign and the IWC, Japan, Norway, Russia, and Iceland resumed whaling of the minke species in the early 1990's, citing their rights to refuse specific IWC rulings. These countries based their decisions on information submitted by the Scientific Committee of the IWC, which stated that the minke population was large enough to absorb sustainable exploitation. However, a deoxyribonucleic acid (DNA) test of whale meat imposed by the IWC in 1995 revealed that Japan was also harvesting some fin and humpback whales. At its May, 1995, meeting, the IWC strongly censured the continued whaling activities of these countries and took serious steps toward stopping pirate whalers and the illegal trade of whale meat.

Alvin K. Benson

FURTHER READING

Heazle, Michael. *Scientific Uncertainty and the Politics of Whaling.* Seattle: University of Washington Press, 2006.

Kalland, Arne. *Unveiling the Whale: Discourses on Whales and Whaling.* New York: Berghahn Books, 2009.

Vaughn, Jacqueline. *Environmental Politics: Domestic and Global Dimensions.* 5th ed. Belmont, Calif.: Thomson/Wadsworth, 2007.

Schumacher, E. F.

CATEGORY: Activism and advocacy
IDENTIFICATION: German British economist
BORN: August 16, 1911; Bonn, Germany
DIED: September 4, 1977; on a train en route to Zurich, Switzerland
SIGNIFICANCE: Schumacher's promotion of nonmaterialist values and his writings emphasizing the importance of protecting resources while attending to the needs of humans had a great influence on the environmental movement in the 1970's and 1980's.

Economist E. F. Schumacher became an important voice in environmental discourse when he published *Small Is Beautiful: Economics As If People Mattered* in 1973. Appearing first in Great Britain, the book found only a small audience in England and Europe, but its U.S. publication in 1974 led to Schumacher's becoming a guru to millions of environmentalists. The work's central theme, that people must replace materialistic values with nonmaterialistic ones, resonated with many Americans, especially college students. By the end of 1974 Schumacher had scheduled hundreds of lecture appearances, most of them in the United States.

Much of what is contained in *Small Is Beautiful* was foreshadowed by developments in Schumacher's life during the 1960's. At the time, Schumacher was the principal economic consultant to the National Coal Board in England. The German-born Schumacher had served as an adviser to British government officials during the 1940's. From 1946 to 1950 he had assisted with economic planning in British-occupied West Germany, and his work there gained the attention of famed economist John Maynard Keynes, who recommended him to the National Coal Board in 1950.

Schumacher remained with the Coal Board for more than twenty years and served as its statistical director from 1963 to 1970. In the midst of this fairly routine career as an expert economist, Schumacher experienced a spiritual crisis that changed the focus of his life. While visiting Burma as an economic consultant in the early 1960's, Schumacher became enchanted with Buddhism. He began to speak of "Buddhist economics," by which he meant the combination of spiritual harmony and material well-being. Although he eventually came to the opinion that Buddhism is not applicable to Western culture (he became a Catholic), the influence of Buddhist philosophy is evident in *Small Is Beautiful*.

The main theme of *Small Is Beautiful* is that modern industrial production, with its emphasis on the most advanced technology, destroys the creativity and dignity of the worker. Schumacher therefore urges the use of intermediate or "appropriate technology" as a way of giving workers a greater sense of satisfaction. He advocates small-scale, regional industry rather than huge national or international corporations.

Throughout *Small Is Beautiful* Schumacher concentrates on the needs of people as opposed to an exclusive concern about the environment. Therefore, he is usually categorized as a social ecologist rather than as a pure or "deep" environmentalist. *Small Is Beautiful* particularly encourages a regional approach to job creation for developing countries. Schumacher argues that such an approach can give a nation's people a sense of achievement and encourage them to protect their resources while gradually increasing the country's pool of skilled workers. In addition, it can slow the drift to urban areas, which leads to greater human misery and more pollution of the air and water. Schumacher's recommendations proved unrealistic for many of the world's less developed nations, but his ideas inspired greater regionalization of industry in the more environmentally conscious industrialized economies during the 1970's and 1980's.

Ronald K. Huch

FURTHER READING

De Steiguer, J. E. "E. F. Schumacher's *Small Is Beautiful*." In *The Origins of Modern Environmental Thought*. Tucson: University of Arizona Press, 2006.

Pearce, Joseph. *Small Is Still Beautiful: Economics As If Families Mattered*. Wilmington, Del.: ISI Books, 2006.

Schumacher, E. F. *Small Is Beautiful: Economics As If People Mattered*. 1973. Reprint. Point Roberts, Wash.: Hartley & Marks, 1999.

Sea Shepherd Conservation Society

CATEGORIES: Organizations and agencies; activism and advocacy; animals and endangered species
IDENTIFICATION: International nonprofit organization devoted to the conservation of marine wildlife

DATE: Founded in 1977

SIGNIFICANCE: The Sea Shepherd Conservation Society's use of direct action and confrontation to disrupt fishing, seal hunting, and whaling operations have been the subject of controversy, but the group's activities have nonetheless focused public and policy attention on overexploitation of the seas.

In 1977 Paul Watson, a founding member of the environmental organization Greenpeace, left that organization after a disagreement with other members and founded the Earthforce Society, the group that later became the Sea Shepherd Conservation Society (SSCS). One of the first interventions carried out by Earthforce was the ramming of a vessel hunting whales in contravention of the worldwide whaling moratorium, an incident that set the tone for subsequent campaigns. Other actions the group has carried out since then have included marking baby seals with an organic dye so their pelts would be worthless to hunters, exposing illegal shark finners, capturing and destroying drift nets, and harassing the Japanese whaling fleet in the Southern Ocean Whale Sanctuary.

SSCS has been controversial since its inception. Detractors often label it a vigilante organization (a label SSCS proudly adopts) or an ecoterrorism group, whereas supporters note that SSCS enforces marine conservation laws and treaties that would otherwise go unenforced. On occasion SSCS has partnered with governments unable to police their own waters for illegal fishing, but it also operates on its own in national waters and on the high seas. SSCS cites Article 21 of the United Nations World Charter for Nature as the legal grounding for its activities, although legal scholars note that the organization's interpretation is probably outside the spirit of the text. Despite efforts by governments of targeted operations to disenfranchise SSCS by impounding its vessels, arresting members, and protesting to the nations in which SSCS branches are located, the organization remains largely unhindered. In fact, it has often used its encounters with law-enforcement agencies and upset citizenry to increase its profile in the news media.

Adam B. Smith

Silent Spring

CATEGORIES: Activism and advocacy; pollutants and toxins

IDENTIFICATION: Book by Rachel Carson that presents an account of the dangers of toxic substances in the environment

DATE: Published in 1962

SIGNIFICANCE: Often credited with helping to launch the modern environmental movement, *Silent Spring* exposed the toxic effects of chemical pesticides on the natural environment. The national debate sparked by the book led to an investigation of pesticide use by the U.S. Congress.

Silent Spring is the best-known work of marine biologist and ecologist Rachel Carson, who is widely considered to be the founder of the research field of environmental ethics. *Silent Spring*, like Carson's many pamphlets and other books on conservation and natural resources, frames an environmental ethics around the holistic thesis that human beings are but a single part of the whole of nature, distinguished primarily by their power to alter the natural world, which all too often they do for the worse and irreversibly.

Carson was motivated to write *Silent Spring* by a re-

Author and environmentalist Rachel Carson. (Library of Congress)

port from friends regarding the broad lethal effects of the aerial spraying of the insecticide dichloro-diphenyl-trichloroethane (DDT) on the wildlife at their bird sanctuary. The book was more than four years in preparation; Carson investigated the effects of chemical pesticides on the chain of natural life, poring over the work of a wealth of researchers, and secured damning evidence of the heedless pesticide poisoning of American air, rivers, and soils. The title of the book derives from its apocalyptic opening chapter, which pictures how the world would look and sound if indiscriminate spraying of pesticides were to continue. With *Silent Spring*, Carson challenged the U.S. government, agricultural scientists, and chemical pesticide producers with evidence of ecological and societal carelessness in the unrestricted use of chemical toxins. She warned the public that pesticides that had been in widespread use since World War II had not been tested to determine their long-term effects.

Silent Spring launches its environmental ethics from the grounding assumption that because human beings are an integral part of the natural world, the largely confrontational approach that humans, and in particular the scientific community, have traditionally assumed toward nature is an inappropriate one, governed by notions of control and manipulation. Since this orientation places humans in a state of war against nature, it simultaneously sets them at war against themselves, as earthly beings. Carson contends that humankind's integral position and remarkable scientific and technological power places on human beings a moral requirement: to approach their relationship with nature as a calling, a moral duty to stewardship.

Responses

When it first appeared in 1962, *Silent Spring* immediately sparked a national debate between opponents and proponents of the use of synthetic chemical pesticides. The public was shocked to learn the extent of the peril associated with the widespread use of pesticides across the country. The response from the chemical industry and the government was to name Carson an alarmist; she was accused of being excessively emotional and lacking in rational objectivity—in short, unscientific. The Monsanto Company, one of the nation's largest chemical companies, launched a counterattack to *Silent Spring* that included commissioning articles that adopted Carson's poetic style but presented a different apocalyptic vision, one in which insects and other pests strip the countryside, leaving the planet uninhabitable by humans.

Despite the virulent counterattacks, Carson's credibility was affirmed in 1963 when she was called as an expert to testify before the U.S. Congress, which was investigating the dangers of pesticide use. In her testimony she repeated the main theses of *Silent Spring*, revealing that since the 1940's the number of chemicals created to control insects, rodents, weeds, and other organisms had proliferated to more than two hundred, and that poisonous sprays, dusts, and aerosols, under thousands of brand names, were being applied universally on farms, in forests, in homes, and in gardens. These chemical toxins are so dangerous for people and the environment, Carson argued, that they should not be called insecticides but biocides, for their ability to poison all earthly life. Carson called on Congress to take decisive action to protect human health and the environment against the toxins.

Even before its publication date, *Silent Spring* had become a best seller, and the following year (1963), when it was published in England, it again reached best-seller status. One of this work's lasting effects has been to bring the topic of environmental ethics into the public sphere. *Silent Spring* gave pause to a generation that had previously trusted the future of the natural and human world to science, the government, and corporate interests. Carson died in 1964, after a long battle against cancer, but the fight she began during the 1950's to preserve the beauty and integrity of earthly life continues to inspire new generations. The fierce debate touched off by *Silent Spring* continues in the twenty-first century among scientists and philosophers as much as among laypersons: whether the purpose of science is to dominate and alter nature to serve human purposes or to preserve nature, study its mysteries, and find the place of humankind within it.

Wendy C. Hamblet

Further Reading

Carson, Rachel. *Silent Spring*. 40th anniversary ed. Boston: Houghton Mifflin, 2002.

Dunlap, Thomas R., ed. *DDT, "Silent Spring," and the Rise of Environmentalism: Classic Texts*. Seattle: University of Washington Press, 2008.

Lytle, Mark Hamilton. *The Gentle Subversive: Rachel Carson, "Silent Spring," and the Rise of the Environmental Movement*. New York: Oxford University Press, 2007.

Murphy, Priscilla Coit. *What a Book Can Do: The Publication and Reception of "Silent Spring."* Amherst: University of Massachusetts Press, 2005.

Silkwood, Karen

CATEGORY: Nuclear power and radiation
IDENTIFICATION: American nuclear industry worker and union activist
BORN: February 19, 1946; Longview, Texas
DIED: November 13, 1974; near Crescent, Oklahoma
SIGNIFICANCE: Until her death in an automobile crash Silkwood was not widely known, but after the accident, which occurred under mysterious circumstances, many antinuclear activists saw her as a martyr to their cause.

Karen Silkwood, the daughter of Bill and Merle Silkwood, grew up in the oil and gas fields around Nederland, Texas. She eloped with Bill Meadows after one year of study at Lamar College, where she was enrolled in a course of study leading to a degree in medical technology. The couple separated in 1972, and their three children remained with their father. Silkwood moved to Oklahoma to be near her parents and found work as a laboratory technician with the Kerr-Magee Corporation at its Cimarron plant in the town of Crescent.

Silkwood joined the Oil, Chemical, and Atomic Workers Union and was soon involved in actions protesting Kerr-Magee's lax health and safety procedures. One allegation she made was that the company had falsified records to cover up the fact that it was missing some nuclear material. Her concern in this area led to her being called to testify before the Atomic Energy Commission in Washington, D.C., in the early part of 1974. At that time Silkwood apparently agreed to clandestinely obtain film evidence of poor workmanship in Kerr-Magee's manufacture of nuclear reactor fuel rods.

In early November of 1974, monitoring equipment in the Kerr-Magee plant detected that Silkwood had become contaminated with a radioisotope. Further testing showed that the contamination was also present in her apartment. Silkwood was taken to the Los Alamos National Laboratory, where more refined testing was performed. It was determined that the level of the contamination was not serious and did not constitute a threat to her health.

On the night of her death, Silkwood left a union meeting supposedly carrying an envelope that contained proof of wrongdoing by Kerr-Magee. She was on her way to a meeting with Drew Stephens, a *New York Times* reporter and union representative, when her car crashed into a culvert alongside a dry, straight section of Oklahoma Highway 74. Silkwood was killed in the crash, and the envelope was never recovered. The Oklahoma Highway Patrol concluded that she had fallen asleep while driving, but a private investigator concluded that her car had been forced off the road.

The resulting controversy, charges of a cover-up, and lawsuits provided a focal point for those concerned with safety in the nuclear power industry. Congressional hearings brought forth intriguing and bizarre stories but led to no definite conclusions. The Atomic Energy Commission confirmed that safety violations had occurred at the Cimarron plant, and it was eventually closed. After years in court, most of the questions surrounding Silkwood's death remained unanswered, but her father did win a large settlement from Kerr-Magee on behalf of her children. Her death and the events leading up to it are portrayed in the award-winning motion picture *Silkwood* (1983).

Kenneth H. Brown

FURTHER READING

Lief, Michael S., H. Mitchell Caldwell, and Benjamin Bycel. "Death by Plutonium: Fallout from Karen Silkwood's Death Brings the Nuclear Industry to Its Knees." In *Ladies and Gentlemen of the Jury: Greatest Closing Arguments in Modern Law*. 1998. Reprint. New York: Charles Scribner's Sons, 2008.

Rashke, Richard. *The Killing of Karen Silkwood: The Story Behind the Kerr-McGee Plutonium Case*. 2d ed. Ithaca, N.Y.: Cornell University Press, 2000.

Singer, Peter

CATEGORIES: Activism and advocacy; animals and endangered species
IDENTIFICATION: Australian philosopher and bioethicist
BORN: July 6, 1946; Melbourne, Victoria, Australia
SIGNIFICANCE: Singer, author of the 1975 book *Animal Liberation* and numerous other works in applied ethics, is considered by many to have launched the modern animal liberation movement.

Peter Singer is the son of Jewish parents who fled Vienna in 1938, after the Nazi occupation of Austria. His father was a coffee salesman and his mother one of the first female graduates of the University of Vienna. They settled in Melbourne, Australia. As a young student, Singer excelled in math and science, but preferred the humanities and eventually studied history, philosophy, and law. He received undergraduate and master's degrees from the University of Melbourne and went on scholarship to University College, Oxford, where he received a bachelor of philosophy degree. He held various positions at Monash University in Melbourne, where he helped establish the Centre for Human Bioethics and served as codirector of the Institute for Ethics and Public Policy. In 1999 he took the position of Ira W. DeCamp Professor of Bioethics at Princeton University. He is also Laureate Professor at the University of Melbourne's Centre for Applied Philosophy and Public Ethics.

The widespread popularity of his 1975 book *Animal Liberation* cemented Singer's reputation as one of the most influential moral philosophers of his generation. He is a utilitarian, which means that for him the rightness of an action is determined solely by its consequences. In assessing consequences, Singer argues that one must give equal consideration to the interests of all beings affected. All sentient beings, he argues, share certain interests, such as experiencing pleasure and avoiding pain. In determining the correct course of action, one must weigh interests according to their strength rather than according to one's affinity for the being that has the interest. According to Singer, humankind's current treatment of nonhuman animals does not reflect an equal consideration of the animals' interests. In fact, it barely considers them at all. Singer contends that humans' poor treatment of nonhuman animals is simply the result of speciesist attitudes that are no more justifiable than racism or sexism. Just as racists violate the principle of equality by favoring members of their own race, speciesists violate the principle of equality by allowing the interests of their own species to outweigh the greater interests of members of other species.

Singer has also discussed the connection between a human-centered ethic and environmental values. Essentially, he argues that environmental policy decisions must give equal consideration to all those affected by an action. Cattle farms may stimulate economic growth for humans but cause suffering and death for cattle, and perhaps loss of habitat for other species. The destruction of rain forests, the building of dams, and other human actions may have consequences that affect members of other species with whom humans share the planet. Singer asserts that the interests of these other species cannot be ignored; he argues that appropriate environmental values give equal consideration to the interests of all sentient beings. Moreover, these interests include the consequences of animal production, such as that required for the meat and egg industries. In addition to the suffering of the animals involved, animal production creates massive amounts of waste, increases the use of fossil fuels, pollutes air and water, and releases greenhouse gases into the atmosphere. According to Singer, intensive animal production is a disaster, and he advocates a vegan diet and lifestyle.

Jonelle DePetro

FURTHER READING

Beers, Diane L. *For the Prevention of Cruelty: The History and Legacy of Animal Rights Activism in the United States.* Athens: Swallow Press/Ohio University Press, 2006.

Schaler, Jeffrey A., ed. *Peter Singer Under Fire.* Chicago: Open Court, 2009.

Singer, Peter. *Animal Liberation.* 1975. Reprint. New York: HarperPerennial, 2009.

_____. *Writings on an Ethical Life.* New York: Ecco Press, 2000.

Snyder, Gary

CATEGORIES: Activism and advocacy; preservation and wilderness issues
IDENTIFICATION: American poet, essayist, and environmental activist
BORN: May 8, 1930; San Francisco, California
SIGNIFICANCE: Snyder was one of the first writers to base his poetry, ethics, and spirituality in environmental ideas and values. He is one of the most influential figures in American nature writing.

Gary Snyder grew up in a rural area outside Seattle, Washington, and started hiking in the mountains early in his life. The radical labor politics of the region in which he lived laid the foundation for his critique of the dominant ideology and social structure of Western society, and the presence of American

Poet and environmental activist Gary Snyder. (Courtesy, author)

Indian and Asian culture in the Pacific Northwest attracted him to other traditions as constructive alternatives. In college he studied anthropology, followed by graduate study in Asian languages. He was a logger, mountain lookout, and merchant seaman; a love for work and an emphasis on the physical are key themes in his writings.

In the 1950's Snyder moved to the San Francisco area, where he was associated with (but not a member of) the Beat movement. His early poetry was particularly influenced by Native American myth and Chinese nature poetry, which he translated into English. He studied Zen Buddhism in Japan during the late 1950's and early 1960's, developing a rich understanding of both the philosophy and practice of that tradition. Following his return to the United States in the late 1960's, Snyder began emphasizing Native American wisdom along with Buddhism in his view of nature and the place of humans within it.

Snyder writes of the sacredness of natural processes, including predation and decomposition. He argues for the intrinsic value of all of nature and the importance of biodiversity, celebrating "the preciousness of mice and weeds." He thus goes beyond the traditional romantic view of nature. He is known not only for his praise of wilderness but also for his exploration of its connection to the rich wilderness in the human mind. His perspective on the self and nature exhibits the Buddhist view of radical interrelationship, which combines a holistic vision of nature with an affirmation of the reality and value of individuals.

Snyder's poetry is suffused with a sensuous and mystical intimacy with nature, and Snyder has proposed a shamanistic view of the poet's role as one who heals by bringing people into a close relationship with the natural world. Yet he also insists on attention to the practical details of living as a full member of one's bioregion. He sees the interplay of culture and nature as central to the development of a deeply rooted sense of place. Snyder is considered a major voice of the deep ecology movement, and his critique of Western society and ideology relates him to social ecology and ecofeminism as well. Among the honors Snyder has received for his work are the Pulitzer Prize in poetry for his collection *Turtle Island* (1974) and the 2008 Ruth Lilly Poetry Prize, which is presented for lifetime achievement.

David Landis Barnhill

FURTHER READING

Hart, George. "Gary Snyder, *Turtle Island* (1974)." In *Literature and the Environment*, edited by George Hart and Scott Slovic. Westport, Conn.: Greenwood Press, 2004.

Snyder, Gary. *A Place in Space: Ethics, Aesthetics, and Watersheds*. 1995. Reprint. Berkeley, Calif.: Counterpoint, 2008.

Sun Day

CATEGORIES: Activism and advocacy; energy and energy use
THE EVENT: Day set aside by the U.S. government to increase awareness of solar energy and encourage the development of solar technologies
DATE: May 3, 1978
SIGNIFICANCE: Sun Day was successful in that it helped policy makers gather input for the develop-

ment of national solar energy policies and raised awareness of the benefits of solar power, but progress in solar technologies continued at a relatively slow pace after the event, and the United States continued to be heavily dependent on fossil fuels.

During the 1960's and 1970's expanding demands for energy, increasing concerns regarding environmental quality, and limited domestic capacity to meet energy demands with traditional fossil fuels brought many Americans to the realization that the United States needed to place a higher priority on renewable sources of energy, particularly solar energy. The urgency of the problem was dramatically impressed upon the leadership of the nation with the Middle East oil embargo in 1973. It became clear to the American public that while the oil embargo would eventually pass, the nation, and even the world, could never again operate under the assumption that the traditional dependence on fossil fuels and other existing sources of energy could continue.

The need for a comprehensive program aimed at developing solar energy as a viable contributor to the future energy supply in the United States led to the creation of the Solar Energy Research Institute (SERI) in Denver, Colorado, in 1977 and the designation of May 3, 1978, as Sun Day. Solar energy awareness and development were emphasized throughout the week of May 1 through May 7, 1978. SERI provided technical support to the federal Sun Day Committee in its efforts to generate large volumes of information on solar energy for the public. A SERI-produced slide show on the technology and potential of solar energy was distributed throughout the nation to be shown at regular intervals in larger cities during the week. At the U.S. Customs House in Bowling Green, New York, solar energy displays were open to public view from May 3 to May 7.

On Sun Day, President Jimmy Carter visited SERI and gave an address on the future of solar energy in the United States in which he requested that every federal government agency consider more ways to help solar energy become a part of everyday American life. Carter pointed out the importance of developing renewable and essentially inexhaustible sources of energy in the future, particularly placing new emphasis on the importance of solar energy in the country's coming energy transition. He concluded that the costs associated with solar power technologies must be reduced so that solar power could be used more widely and would help establish a cap on rising fossil-fuel prices. In addition, Carter stated that he had just provided the U.S. Department of Energy with an additional $100 million for expanded efforts in solar research, development, and demonstration projects.

Following Carter's Sun Day address, a series of well-attended forums were conducted across the country. Participants included congressional representatives; state and local government officials; representatives of industries, labor organizations, public utilities, and special interest groups; and members of the general public. These public forums identified citizen groups interested in solar energy and provided input for the development of national solar energy policies.

Alvin K. Benson

FURTHER READING

Laird, Frank N. *Solar Energy, Technology Policy, and Institutional Values*. New York: Cambridge University Press, 2001.

Scheer, Hermann. *A Solar Manifesto*. 2d ed. London: James & James, 2001.

Union of Concerned Scientists

CATEGORIES: Organizations and agencies; activism and advocacy; human health and the environment

IDENTIFICATION: American nonprofit organization devoted to advocacy for science and opposition to the misuse of science and technology

DATE: Established in 1969

SIGNIFICANCE: The Union of Concerned Scientists has raised public awareness and influenced policy makers regarding many environmental issues with international impact, including global climate change, nuclear arms control, and agricultural biotechnology.

The Union of Concerned Scientists (UCS) was created during the Vietnam War era when faculty and students at the Massachusetts Institute of Technology (MIT) organized to call for the application of scientific research to pressing environmental and social problems rather than to military programs. Months of planning and preparation by eminent MIT scientists led to a voluntary research stoppage and day of education and discussion on March 4, 1969. News of the planned meetings at MIT spread to other universities, and meetings and protests took place on campuses elsewhere that same day.

UCS has since expanded its interests beyond protesting military research, and its membership has grown to become an alliance of scientists and citizens across the United States; by 2008 the organization had more than 78,000 members. It is headquartered in Cambridge, Massachusetts, and has offices in Washington, D.C.; Berkeley, California; and Chicago, Illinois.

UCS combines scientific research with legislative advocacy, grassroots organizing, and media coverage aimed at promoting a safer and healthier environment. A core group of researchers and policy experts collaborates with others to provide credible scientific information for use in citizen advocacy, consumer education, and expert testimony. The organization helps to coordinate and lobby for diverse projects on such issues and concerns as nuclear power safety, renewable energy sources, arms control, global climate change, international family-planning programs, agricultural biotechnology safety, clean vehicles, sustainable transportation policies, and scientific integrity.

UCS advocacy has included support for a plan that would advance the prospects for renewable energy under electric utility deregulation by increasing requirements for dollars spent on research for renewable energy sources by utilities and establishing a trust fund to support this research and development. UCS research and advocacy have also supported efforts to reduce reliance on single-passenger cars and provide incentives for the manufacture and purchase of cleaner vehicles.

U.S. Environmental Protection Agency policy regarding genetically engineered foods was influenced significantly by the book *The Ecological Risks of Engineered Crops* (1996), a revised and expanded version of a UCS report authored by Jane Rissler and Margaret Mellon. UCS has also worked on the issue of family planning on an international level and has helped convince the U.S. Congress to restore funding for family-planning programs. In addition, UCS has waged campaigns to pressure Congress and the president of the United States to take part in such international agreements as the Chemical Weapons Convention, the Comprehensive Nuclear-Test-Ban Treaty, and the Strategic Arms Reduction Treaty.

In 1992, a few months after the Earth Summit was held in Rio de Janeiro, Brazil, UCS issued a document titled "World Scientists' Warning to Humanity." More than 1,700 scientists signed the declaration, which calls for scientists, governments, business and industry, and citizens to work cooperatively to minimize carbon dioxide emissions and other stresses that human activity places on the ecosphere. In 1997 UCS organized the Science Summit on Climate Change, an event that clarified a growing consensus among scientists about global warming. This summit produced "World Scientists' Call for Action," a petition signed by 1,586 scientists (including 110 Nobel laureates) from sixty-three countries. The petition, which calls for an effective international climate treaty, helped shape the Kyoto Protocol. UCS has since issued a number of publications on greenhouse gas emissions, the possible impacts of projected climate change, and recommended emission reduction efforts.

During President George W. Bush's administration (2000-2008), UCS published reports claiming that scientific findings within several U.S. government agencies were being manipulated, suppressed, and distorted. UCS shared its recommendations for reversing the politicization of science with President Barack Obama's transition team, and in early 2009 the new president directed the Office of Science and Technology Policy to create a plan to ensure scientific integrity throughout the executive branch.

Anne Statham
Updated by Karen N. Kähler

FURTHER READING

Cleetus, Rachel, Steven Clemmer, and David Friedman. *Climate 2030: National Blueprint for a Clean Energy Economy.* Cambridge, Mass.: Union of Concerned Scientists, 2009.

Kendall, Henry. "Union of Concerned Scientists' Warning." In *Life Stories: World-Renowned Scientists Reflect on Their Lives and the Future of Life on Earth*, edited by Heather Newbold. Berkeley: University of California Press, 2000.

Moore, Kelly. *Disrupting Science: Social Movements, American Scientists, and the Politics of the Military, 1945-1975.* Princeton, N.J.: Princeton University Press, 2008.

Rissler, Jane, and Margaret Mellon. *The Ecological Risks of Engineered Crops.* Cambridge, Mass.: MIT Press, 1996.

U.S. Climate Action Partnership

CATEGORIES: Organizations and agencies; activism and advocacy; weather and climate
IDENTIFICATION: Alliance of business corporations and leading environmental organizations that promotes the passage of federal legislation to reduce greenhouse gases
DATE: Established on January 22, 2007
SIGNIFICANCE: The U.S. Climate Action Partnership represents a shift in direction among many corporations toward working with environmentalists and others to address the problem of global warming.

Members of the U.S. Climate Action Partnership (USCAP) pledge to work with the Congress and the president of the United States, as well as other stakeholders, in enacting climate protection programs that are fair, economically sustainable, and environmentally effective, with the goal of significantly reducing greenhouse gas emissions. Fourteen businesses and four environmental organizations were the initial founding members of USCAP. Since its founding, the organization has both gained and lost members; environmental organizations that were members in 2010 included the Environmental Defense Fund, the Natural Resources Defense Council, and the Nature Conservancy, and business members included Alcoa, Chrysler, the Ford Motor Company, PepsiCo, and the Dow Chemical Company.

USCAP outlines a series of design principles and recommendations for legislation to address the problem of global warming in a document titled *A Call for Action*. In 2009 USCAP published *A Blueprint for Legislative Action: Consensus Recommendations for U.S. Climate Protection Legislation*, which addresses how the United States can reduce greenhouse gas emissions without hindering the strength and productivity of the American economy. Among the recommendations offered in this document are that businesses be allowed to invest in forest conservation programs to meet their required emission reductions and that businesses be offered incentives to engage in the development and deployment of emissions-reducing technologies.

USCAP asserts that policy changes based on its recommendations would result in innovation, energy security, and economic growth. Furthermore, such changes would improve the U.S. balance of trade and demonstrate U.S. leadership on the issue of climate change. Critics have argued, however, that USCAP supports self-serving regulations and policy changes that would be beneficial to its members, such as a cap-and-trade system to reduce carbon dioxide emissions; many environmentalists assert that cap-and-trade systems in effect reward polluters by protecting them from paying the cost of compliance with regulations.

Prior to joining USCAP, some of the group's members were strongly opposed to any regulation aimed at reducing greenhouse gases and lobbied against U.S. ratification of the Kyoto Protocol. Some scientists and environmentalists have expressed concerns that the same companies that are members of USCAP are also engaged in lobbying against USCAP's stated goals.

Lakhdar Boukerrou

Watson, Paul

CATEGORIES: Activism and advocacy; animals and endangered species
IDENTIFICATION: Canadian animal rights and environmental activist
BORN: December 2, 1950; Toronto, Ontario, Canada
SIGNIFICANCE: A dissident Greenpeace member and experienced sailor, Watson founded the Sea Shepherd Conservation Society, one of the world's most aggressive environmental organizations. He and his organization have mounted vigilante (but deliberately nonlethal) attacks against the efforts of seal hunters, whalers, and drift-net fishers.

A onetime officer in the Canadian Coast Guard, Paul Watson helped to found Greenpeace in 1971. By the mid-1970's, however, he found himself increasingly at odds with the group's philosophy. Initially a small band of dedicated activists who placed themselves at personal risk to interfere with such environmentally negative practices as nuclear weapons tests, Greenpeace had grown into an international concern that directed much of its effort toward fundraising. Watson believed that the group's evolution had dissipated its effectiveness, and he also chafed at the organization's commitment to passive protest in pursuit of its goals.

Watson believed that doing harm to living things is wrong, but the use of nonharmful force against property could be justified to protect life. Moreover, he became increasingly convinced that passive protest was not working. In March, 1977, while participating in a Greenpeace protest of the slaughter of baby harp seals by hunters using clubs in Canada's Gulf of St. Lawrence, Watson seized a club from a seal hunter and threw it into the water. Such direct action conflicted with Greenpeace's expressed policy, and he was expelled from the group. When Greenpeace's board of directors charged Watson with vigilantism, he replied that, in the absence of an environmental police force to oppose environmental crimes, environmental vigilantes were bound to appear. Greenpeace soon distanced itself from Watson and his unapologetically extreme methods.

In the summer of 1977 Watson and several friends established an organization they at first called Earthforce; it later became the Sea Shepherd Conservation Society. Headquartered in Vancouver, Canada, the group was dedicated to the use of direct action to protect the world's animals. Although Watson and his band of activists would later achieve notoriety for defending marine life, on their first mission they traveled to East Africa to document the killing of elephants for ivory. Earthforce presented its findings, including film of the illegal slaughter of elephants, to the U.S. government in support of a legislative ban on the importation of African ivory.

For Watson, however, the preservation of marine mammals was a recurring goal. With the aid of a grant from the writer and animal rights activist Cleveland Amory, Watson purchased a retired fishing boat, which he christened the *Sea Shepherd*, and hired a crew. On the *Sea Shepherd*'s first voyage, Watson and the crew took the ship through some 644 kilometers (400 miles) of ice to the Gulf of St. Lawrence. There they sprayed hundreds of baby seals with a harmless red dye that made the animals' white pelts valueless to hunters. Soon, however, they were stopped and arrested by Canadian police. Undeterred, Watson and his allies would return to the gulf in 1982 and 1983, after which the gulf hunt was discontinued.

Aggressive Tactics

Publicity stemming from the group's sometimes violent encounters with sealers and from its brushes with the law had led to steady membership increases, and the scope and ambition of the group's activities expanded as well. In the spring and summer of early 1979, Watson and other members of the Sea Shepherd Conservation Society organized what was essentially an espionage ring that tracked the movements of the *Sierra*, a notorious "pirate" whaling ship whose crew had killed an estimated four hundred whales per year since the 1960's. In July, 1979, Watson and two assistants aboard the *Sea Shepherd* located the *Sierra* off the coast of North Africa and followed the whaling ship to the harbor in Leixões, Portugal. Outside the port, the larger and faster *Sea Shepherd* twice rammed the *Sierra*, rupturing the whaler's side but causing no injuries to the crew. The Portuguese navy took the *Sea Shepherd* into custody, but the *Sierra* had incurred enormous damage and had to be towed to Lisbon for repairs. After Portuguese authorities threatened to turn the *Sea Shepherd* over to the *Sierra*'s owners as compensation, Watson and two friends sneaked into the Lisbon harbor and scuttled the ship by opening its pipes.

Watson and the Sea Shepherd Conservation Society undertook similar actions many times in ensuing years, actively confronting whalers, sealers, and driftnet fishers, attacking their vessels and equipment, and often ending up in court. By the 1990's, the organization had grown to include thousands of members and was operating a fleet of antiwhaling vessels. It continues its radical practices in the early twenty-first century and remains a source of controversy.

Watson is a professional lecturer and the author of several books on protecting wildlife and marine life. In 2000 *Time* magazine named him one of the environmental heroes of the twentieth century. He is the subject of the 2008 biographical film *Pirate for the Sea* and has appeared in a number of other documentaries. He is also a central figure in *Whale Wars*, a television reality

---

The Sea Shepherd Conservation Society's Mission

The Sea Shepherd Conservation Society's mission statement reads as follows:

Established in 1977, Sea Shepherd Conservation Society (SSCS) is an international non-profit, marine wildlife conservation organization. Our mission is to end the destruction of habitat and slaughter of wildlife in the world's oceans in order to conserve and protect ecosystems and species.

Sea Shepherd uses innovative direct-action tactics to investigate, document, and take action when necessary to expose and confront illegal activities on the high seas. By safeguarding the biodiversity of our delicately-balanced ocean ecosystems, Sea Shepherd works to ensure their survival for future generations.

series that debuted on the Animal Planet channel in 2008. The program documents the exploits of Watson and the Sea Shepherd fleet as they fight to deter Japanese whaling vessels hunting off the Antarctic coast.

Opponents of Watson's extreme methods have portrayed him and his followers as fanatics and have accused them of using terrorist tactics. Watson has countered that he and his organization have never injured a human being, committed an act of violence against a living creature, been convicted of a felony, or been sued. He has maintained that he does not break the law, but rather upholds it, and he has condemned society's failure to save imperiled creatures from slaughter as being in itself an act of violence.

Robert McClenaghan
Updated by Karen N. Kähler

Further Reading

Heller, Peter. *The Whale Warriors: The Battle at the Bottom of the World to Save the Planet's Largest Mammals.* New York: Free Press, 2007.

Morris, David B. *Earth Warrior: Overboard with Paul Watson and the Sea Shepherd Conservation Society.* Golden, Colo.: Fulcrum, 1995.

Watson, Paul. *Ocean Warrior: My Battle to End the Illegal Slaughter of Marine Life on the High Seas.* London: Vision, 2003.

_____. *Seal Wars: Twenty-five Years on the Front Lines with the Harp Seals.* Buffalo, N.Y.: Firefly Books, 2003.

Watson, Paul, Warren Rogers, and Joseph Newman. *Sea Shepherd: My Fight for Whales and Seals.* New York: W. W. Norton, 1982.

Weyler, Rex. *Greenpeace: How a Group of Ecologists, Journalists, and Visionaries Changed the World.* Emmaus, Pa.: Rodale Press, 2004.

White, Lynn Townsend, Jr.

Category: Activism and advocacy
Identification: American historian and author
Born: April 29, 1907; San Francisco, California
Died: March 30, 1987; Los Angeles, California
Significance: White argued that religion—medieval Christianity, in particular—played a significant role in the environmental crisis that was becoming apparent during the late 1960's. His controversial thesis was influential in spawning several movements in environmentalism, including ecotheology.

Lynn Townsend White, Jr., was a professor of history at Princeton and Stanford universities and at the University of California, Los Angeles. A graduate of Stanford, Union Theological Seminary, and Harvard University, he also served as president of Mills College and was a founding member of the Society for the History of Technology.

As a historian, White specialized in the history of medieval technology. His major work, *Medieval Technology and Social Change*, published in 1962, defined the field of medieval technology historiography. The book details several seemingly small changes in agriculture and animal husbandry—the stirrup, the plow, and crop rotation, for example—that profoundly influenced European culture and land use.

In regard to environmental issues, White is best remembered for his seminal 1966 lecture "The Historical Roots of Our Ecological Crisis," which was published in 1967 in the journal *Science*. In this work White argues that ecology is directly related to human beings' beliefs about themselves and about the world. According to White, Western Christianity promotes the belief that nature exists only to serve humanity, and this idea leads to the ruthless exploitation of nature.

The article has frequently been taken as an attack on Christianity, but White, a lifelong Presbyterian, did not intend to attack Christianity as a whole; he simply sought to urge a reexamination of Christianity's doctrine of nature. Against the hierarchical view that humanity should dominate nature, White's essay proposes Saint Francis of Assisi as the "patron saint for ecologists" because Francis's theology regards all natural things as having been made for the glory of their creator and therefore as intrinsically valuable.

White's thesis is not universally accepted by historians, theologians, or ecologists. Nevertheless, its importance and its influence, especially on religious movements such as ecotheology, are broadly recognized.

David L. O'Hara

World Resources Institute

Categories: Organizations and agencies; activism and advocacy; resources and resource management
Identification: Nonprofit organization devoted to protecting the environment by supporting sustainable management of the world's resources

DATE: Established on June 3, 1982

SIGNIFICANCE: Through its many programs, the World Resources Institute works to develop international agreements to protect the environment and promotes responsible investments in environment-friendly energy and transportation technologies.

The World Resources Institute (WRI) was launched in 1982 by James Gustave Speth with a $15 million grant from the John D. and Catherine T. MacArthur Foundation and $10 million seed money from the Andrew K. Mellon Foundation and the Rockefeller Foundation. WRI was founded to be a center that would address policy research on and analysis of global resources, but it has since shifted its mission and goals. The stated mission of the World Resources Institute is to "move human society to live in ways that protect Earth's environment and its capacity to provide for the needs and aspirations of current and future generations." A good portion of WRI funding comes from U.S. government agencies, including the Environmental Protection Agency (EPA), the National Aeronautics and Space Administration (NASA), the U.S. Agency for International Development (USAID), and the Departments of Agriculture and Energy.

WRI program goals are centered on four primary areas: climate protection, governance, markets and enterprise, and people and ecosystems. The organization seeks to be a catalyst in the development of international agreements and U.S. policies to protect the environment, to promote responsible investments in energy and transportation technologies that are friendly to the environment, and to help reduce greenhouse gas emissions through clean alternatives supported by all stakeholders. More specifically, WRI works to empower people and institutions to do the following: mitigate climate change (particularly by reducing greenhouse gas emissions) and adapt to climate change as needed; make informed, socially equitable decisions to ensure environmental sustainability; expand economic opportunities while protecting the environment and harnessing markets; and stop and reverse land degradation and other environmental problems to ensure productivity for generations to come.

WRI sponsors and supports research, workshops, conferences, and related activities to address the world's environmental problems. In addition, WRI produces a number of publications and offers online resources, some of the most important of which are *World Resources* (launched in 1986), a biennial country-by-country assessment of environmental conditions and trends; Global Forest Watch (launched in 2000), an independent online network for monitoring forests; and Earth Trends (launched in 2001), a comprehensive online database of information on the world's social, economic, and environmental trends. The NextBillion.net, an online initiative launched by WRI in 2004, is the result of the first conference held for the purpose of finding ways in which the business community can help alleviate and reduce global poverty by addressing and meeting the needs of the world's poorest people at the "base of the economic pyramid."

WRI played an important role in the creation of the Global Environment Facility (GEF) and the adoption of the United Nations Convention on Biological Diversity (CBD), as well as many other international environment-related efforts. Over the years, WRI has joined with other nonprofit organizations as well as with international agencies and other institutions to address the protection of the world's natural resources. WRI seeks to ensure that decisions made about the use and management of natural resources and ecosystems reflect the needs and values of the people who are dependent on those resources and ecosystems.

WRI developed the Greenhouse Gas Protocol Initiative in collaboration with the World Business Council for Sustainable Development. This comprehensive tool provides accounting information and information on standards on nearly every greenhouse gas program in place in the world. WRI is also a founding member and supporter of the U.S. Climate Action Partnership (USCAP), an alliance of businesses and environmental groups that promotes the enactment of U.S. legislative action on the reduction of greenhouse emissions. The "blueprint for legislative action" developed by USCAP members and issued on January 15, 2009, supports the implementation of a cap-and-trade system, investment in carbon capture and storage technology, and free allowances for certain businesses.

Lakhdar Boukerrou

FURTHER READING

Irwin, Frances, et al. *Restoring Nature's Capital: An Action Agenda to Sustain Ecosystem Services.* Washington, D.C.: World Resources Institute, 2007.

Speth, James Gustave. *Red Sky at Morning: America and the Crisis of the Global Environment.* New Haven, Conn.: Yale University Press, 2004.

World Trade Organization

CATEGORIES: Organizations and agencies; resources and resource management
IDENTIFICATION: International organization that deals with the rules of trade between nations
DATE: Established on January 1, 1995
SIGNIFICANCE: The World Trade Organization was created to regulate global trade among member states and to promote trade liberalization. Its primary goal has been to remove barriers to trade, and because national policies concerning environmental protections are often seen as obstructing free trade, the organization's rulings have frequently failed to place a high priority on such protections.

The World Trade Organization (WTO) administers trade between nations in accordance with sets of rules known as the WTO agreements, which are the results of negotiations among the bulk of the world's trading nations, ratified by the members' governments. The WTO was created in 1995 to ensure that global trade runs smoothly, predictably, and as free of obstructions as possible. It seeks to accomplish this overriding goal by providing a forum for the negotiation of agreements among producers of goods and services, importers, and exporters, granting to these parties a legal and institutional framework for implementing their agreements, monitoring adherence to agreements, and resolving disputes arising from conflicting interpretations and applications of the agreements.

WTO trade agreements are generally reached by consensus among the entire membership (multilateral agreements), but occasionally individual countries use the WTO forum to develop agreements of more limited scope (plurilateral agreements). By 2010 the body of WTO trade agreements consisted of sixteen multilateral agreements and two plurilateral agreements.

By 2010 the WTO had 153 members, 117 of which were developing countries, and was overseeing more than 97 percent of all global trade. The highest body within the organization is the Ministerial Conference, which meets every two years. Between the Ministerial Conference meetings, a General Council conducts WTO affairs. This council meets in Geneva several times each year and also meets as the Trade Policy Review Body and the Dispute Settlement Body. The Goods Council, Services Council, and Intellectual Property Council report to the General Council. Specialized subsidiary bodies administer trade agreements and monitor their implementation and also deal with individual agreements and other areas of negotiation such as the environment, development, membership applications, and regional trade agreements. The WTO is supported by a secretariat, located in Geneva, that is led by a director-general and employs about seven hundred support staff.

ORIGINS

The WTO continues the work of its predecessor, the General Agreement on Tariffs and Trade (GATT), which was created by the victorious Allied Powers in the wake of World War II to regulate international trade to the best advantage of the founding partners. The system was developed over fifty years through a series of trade negotiation rounds. The earliest rounds sought mainly to reduce tariffs, but later negotiations included other nontariff measures, such as antidumping rules. The round of negotiations that took place from 1986 to 1994, known as the Uruguay Round, led to the creation of the WTO.

Further rounds reached agreement on telecommunications services and broadened liberalization measures well beyond those established in the Uruguay Round. By 1994 forty governments had concluded negotiations for tariff-free trade in information technology products, and seventy members reached a financial services deal covering more than 95 percent of trade in banking and other financial services. In 2000 talks began to address the liberalization of trade in agriculture and related services. The fourth Ministerial Conference, held in Doha, Qatar, in November, 2001, incorporated these changes into the WTO agenda. WTO rules have since come to cover antidumping and subsidies, investment, competition policy, trade facilitation, transparency in government procurement, intellectual property, and a range of broader issues raised by developing countries as they struggle to implement existing WTO agreements.

MISSION AND ACTIVITIES

For the most part, the WTO continues the GATT mission of reducing or eliminating obstacles to trade, particularly import tariffs, and governing the rules and conduct of international trade (for example, setting common product standards, monitoring the use of subsidies, and regulating trade-related intellectual property rights). The WTO monitors member adherence to regional and bilateral trade agreements, ensuring transparency in trade activities and intervening to settle member disputes about agreements as they arise. The WTO's mission includes helping developing member countries to build trade capacity and encouraging nonmember countries in the process of becoming members. Other WTO activities include economic research and educating the public about international trade.

Controversies

The WTO asserts that its open-border approach ensures a level playing field among international trading parties, which promotes economic growth in wealthy nations as much as in the developing world. Certainly, GATT and the WTO have helped to build a strong and prosperous trading system that can boast unprecedented economic growth. However, critics argue that the founding and guiding principles of free trade guarantee the most-favored-nation principle, which serves best the interests of the already economically successful while undermining sustainable development and social progress in developing countries, as well as global peace and stability.

Critics note how much global trade rules under the WTO fall short of the institutional framework of the proposed International Trade Organization that was developed following World War II; this framework was designed to balance two paramount goals—the economic goal of global trade liberalization and the social goal of stimulating full global employment. These two goals were to be served simultaneously, not through the enforcement of blanket open-border regulations across the globe but through the granting of special treatment to developing nations, many of them newly emerging from colonial histories. The policy of giving poorer nations a leg up into the global economy was abandoned, however, when the U.S. Senate rejected the broad mandate of the proposed framework. The United States instead led the world's foremost nations in approving the much more limited GATT proposal, which focused primarily on reducing barriers to trade and investment and opening all borders as wide as possible while neglecting the goal of proactive development for poorer nations. During the 1980's, the U.S. government led the charge to replace GATT with a larger, more powerful organization better equipped to carry out that primary task of breaking down barriers to trade. Negotiations concluded in 1994, and the World Trade Organization came into being in 1995.

Wendy C. Hamblet

Further Reading

Anderson, Sarah, and John Cavanagh, with Thea Lee. *Field Guide to the Global Economy*. Rev. ed. New York: New Press, 2005.

Narlikar, Amrita. *The World Trade Organization: A Very Short Introduction*. New York: Oxford University Press, 2005.

Speth, James Gustave, and Peter M. Haas. "Key Actors, Expanding Roles: The United Nations, International Organizations, and Civil Society." In *Global Environmental Governance*. Washington, D.C.: Island Press, 2006.

Wallach, Lori, and Patrick Woodall. *Whose Trade Organization? A Comprehensive Guide to the WTO*. Rev. ed. New York: New Press, 2004.

World Wilderness Congresses

CATEGORIES: Activism and advocacy; preservation and wilderness issues

IDENTIFICATION: International assemblies devoted to the preservation of wilderness on a global scale

DATE: Inaugurated in October, 1977

SIGNIFICANCE: The World Wilderness Congresses have played a major role in wilderness preservation by providing a forum where international participants from diverse cultures and disciplines can address the implementation of global initiatives aimed at preserving the world's biodiversity and its remaining wilderness areas.

The first World Wilderness Congress began as one of the initiatives undertaken by Ian Player, a South African conservationist, to promote wilderness conservation on a global scale and to ensure wilderness preservation. In 1974, convinced that wilderness conservation must be a concern of every nation and every individual, not just a preoccupation of Western culture, Player resigned from the South African Wildlife Service to pursue efforts to make wilderness conservation a global issue. He established the Wilderness Leadership School to introduce individuals from all walks of life to the wilderness by taking them on foot into the African wilderness. Shortly thereafter he founded the WILD Foundation, the Wilderness Trust, and the Magqubu Ntombela Foundation, which honors his friend and colleague Qumbu Magqubu Ntombela, a Zulu chief.

In October, 1977, working with colleagues in conservation, Player and the WILD Foundation convened the first World Wilderness Congress in Johannesburg, South Africa. The congress provided a forum for discussion and implementation of programs and projects targeting wilderness preservation. The focus of the congress was wilderness preservation on an international, multicultural, multiprofessional, and multi-interest scale. Participants in the congress addressed wilderness preservation from many different viewpoints, including artistic, cultural, governmental, environmental, economic, and academic. The organizers emphasized the congress's importance as an opportunity for the exchange of information—not only among the various groups engaged in wilderness preservation but also among those involved in activities affecting wilderness. The congress introduced programs for involving native peoples in wilderness conservation and addressed banking and economic development as wilderness preservation issues. It also recognized the importance of art and creativity in the fight to preserve wilderness by presenting an extensive exhibition of conservation art.

In keeping with its concept of wilderness as a global issue, the WILD Foundation cohosts a World Wilderness Congress every three to five years to provide opportunities for face-to-face discussion and networking among conservation groups from all over the world. The congresses have been held in many different locations, including in Australia, Scotland, Norway, Mexico, India, and Alaska. Significant contributions to wilderness preservation have come out of the congresses, among them a strengthening of the concept that preservation of wilderness is a task for the global community.

Projects and Initiatives

At the second World Wilderness Congress, the issue of the impact of hydroelectric dams on wilderness conservation in Tasmania received international attention for the first time, and participants formulated a global overview of definitions relating to wilderness. At the third congress participants witnessed the effectiveness of gaining global attention for wilderness preservation as it was announced that the government of Tasmania had opted to protect the Southwest Tasmanian Wilderness rather than to build dams. In addition, two new wilderness preservation organizations, Wilderness Association Italiana and the South African Wilderness Action Group, were formed as a result of the third congress.

Two proposals for programs affecting conservation and preservation were made at the fourth congress in 1987; they called for a world conservation bank and a world conservation corps. In 1991 the Global Environment Facility was jointly established by the United Nations Development Programme, the United Nations Environment Programme, and the World Bank as an independent organization that would give financial assistance in the form of grants to developing countries for projects beneficial to the global environment.

The fifth World Wilderness Congress, held in Norway in 1993, focused on preservation of the Arctic in harmony with sustainable use by the indigenous peoples of the region. The sixth congress established the Asian Wilderness Initiative and gathered support for a joint plan of India and Namibia to return cheetahs to India. The seventh congress concentrated on projects and initiatives to increase protected areas of wilderness on both public and private lands and to implement additional legislation and training programs to guarantee wilderness preservation.

The eighth congress, held in Alaska in 2005, gave priority to issues of global warming and to the topic of the petroleum industry's efforts to gain permission to drill for oil and gas in the Arctic National Wildlife Refuge. The congress also saw the creation of the WILD Planet Fund to maintain wilderness and two new organizations: the Native Lands and Wilderness Council, focusing on the participation of indigenous tribes in wilderness preservation through sound land use and management practices; and the International League of Conservation Photographers, an organization of photographers dedicated to increasing appreciation of wilderness and its preservation through photography.

The ninth World Wilderness Congress was held in Mérida, Mexico, in 2009. Participants in the congress addressed the interrelatedness of human activity, wilderness, and climate change. One outcome of the congress was the Message of Mérida, a plan for making the preservation of wilderness and biodiversity a part of the global strategies used to address climate change and its effects. During the congress the WILD Foundation instigated an international agreement for wilderness preservation that was signed by the United States, Canada, and Mexico. In addition, the first Corporate Commitment to Wilderness was formulated and signed by fifteen corporations, with the expectation of more corporations signing in the near future. The ninth congress also produced the Marine Wilderness Collaborative and established a special program for California's marine wilderness.

At each congress, exhibitions of local art, dance, and music are integrated into the programs and recognized as an essential part of wilderness preservation. From the exhibition of conservation art at the first congress to the nature-related aboriginal art presented in Australia, to the twenty life-size jaguar sculptures displayed and the body painting offered at Mérida, the World Wilderness Congresses use art to reiterate the interrelatedness of all peoples and the necessity of wilderness preservation as part of protecting the earth that they share.

Shawncey Webb

FURTHER READING

Chester, Charles C. *Conservation Across Borders: Biodiversity in an Interdependent World*. Washington, D.C.: Island Press, 2006.

Martin, Vance, and Andrew Muir, eds. *Wilderness and Human Communities: The Spirit of the Twenty-first Century*. Golden, Colo.: Fulcrum, 2004.

Martin, Vance, and Partha Sarathy, eds. *Wilderness and Humanity: The Global Issue*. Golden, Colo.: Fulcrum, 2001.

Player, Ian. *Zulu Wilderness: Shadow and Soul*. Golden, Colo.: Fulcrum, 1998.

Worldwatch Institute

CATEGORIES: Organizations and agencies; activism and advocacy; resources and resource management

IDENTIFICATION: Independent research organization that focuses on critical global issues

DATE: Established in 1974

SIGNIFICANCE: The Worldwatch Institute provides important information about the environment to world leaders, policy makers, and the public in general. Through its fact-based analyses and the ideas it offers, the institute helps to shape world opinion regarding environmental protection, development, and sustainable use of resources.

The Worldwatch Institute, the first research institute devoted to the analysis of global environmental issues, was founded in 1974 by Lester Brown, one of the world's most influential thinkers. Based in Washington, D.C., the institute is an interdisciplinary research organization that aims to help create an environmentally sustainable society that is capable of adequately meeting human needs. For this purpose, it focuses on some of the twenty-first century's most pressing global challenges: climate change, resource degradation, population growth, and poverty.

The Worldwatch Institute uses the best available scientific evidence to perform analyses that help shape the views and positions of decision makers and leaders around the world. It aims to promote the development of innovative solutions to global problems by bringing together the public and private sectors as well as concerned citizens. The institute works with a global network of 150 partners and affiliates in forty countries and produces publications, including most notably the influential yearly report titled *State of the World*, that are translated into thirty-six languages.

The main programs of the Worldwatch Institute are the Climate and Energy, Food and Agriculture, Green Economy, China, India, and Transforming Cultures programs. These different programs are described as follows on the institute's dedicated Web site:

- The Climate and Energy Program is dedicated to accelerating the transition to a low-carbon energy system based on sustainable use of renewable energy resources in concert with major energy-efficient gains.
- The Food and Agriculture Program highlights the benefits to farmers, consumers, and ecosystems that can flow from food systems that are flexible enough to deal with shifting weather patterns, productive enough to meet the needs of expanding populations, and accessible enough to support rural communities.
- The Green Economy Program recognizes that the global environmental and economic crises have common origins and must be tackled together. The program seeks to offer solutions that enhance hu-

man well-being and reduce inequities while protecting the planet.
- The China Program seeks to help decision makers within China and around the globe better understand environmental challenges and opportunities.
- The India Program tracks key developments in India and seeks to engage today's decision makers and tomorrow's leaders on all national and global issues.
- The Transforming Cultures Program seeks to transform today's consumerist culture into a culture of sustainability.

In addition to these program areas, the Worldwatch Institute monitors human health, water resources, biodiversity, governance, and environmental security around the world.

The Worldwatch Institute's numerous publications are intended to enable decision makers in government, civil society, business, and academia to keep track of the latest developments concerning the environment and issues of sustainability. The institute's most significant publication, the annual *State of the World*, essentially provides an assessment of global environmental problems and of the innovative ideas proposed and applied across the globe to address them. Every year this report has a particular focus; for example, the 2010 volume is subtitled *Transforming Cultures: From Consumerism to Sustainability*. The 2009 volume focuses on global warming, 2008's examines "innovations for a sustainable economy," and 2007's addresses the "urban future." The institute offers a "*State of the World* at a Glance" feature on its Web site that provides lists and brief explanations of the key facts and innovations noted in each *State of the World* volume.

Among its many other publications, the institute produces *Vital Signs Online*, which provides the latest data and analyses necessary for an understanding of critical global trends, including population growth, biodiversity loss, growth in energy consumption, and rising carbon emissions. *World Watch Magazine*, published bimonthly, offers cutting-edge analysis of social and environmental issues.

Nader N. Chokr

FURTHER READING

Brown, Lester. "Worldwatch." In *Life Stories: World-Renowned Scientists Reflect on Their Lives and on the Future of Life on Earth*, edited by Heather Newbold. Berkeley: University of California Press, 2000.

Nelson, David E. "In Praise of Lester Brown." *Futurist* 42, no. 6 (2008).

Wallis, Victor. "Lester Brown, the Worldwatch Institute, and the Dilemmas of Technocratic Revolution." *Organization and Environment* 10, no. 2 (1997): 109-125.

Worldwatch Institute. *State of the World*. New York: W. W. Norton, 2010.

Zahniser, Howard Clinton

CATEGORIES: Activism and advocacy; preservation and wilderness issues
IDENTIFICATION: American conservationist and nature writer
BORN: February 25, 1906; Franklin, Pennsylvania
DIED: May 5, 1964; Hyattsville, Maryland
SIGNIFICANCE: Zahniser was an influential figure in the wilderness preservation movement of the mid-twentieth century. In addition to serving as executive secretary of the Wilderness Society for more than twenty years, he authored the landmark Wilderness Act of 1964.

Although born in the town of Franklin in northwestern Pennsylvania, Howard Clinton Zahniser grew up to the east in Forest County, in a remote village nestled against the Great Allegheny Forest. After a childhood spent happily roaming the woods and an adolescence during which he absorbed the manifestos of the American Transcendentalists, such as Ralph Waldo Emerson and Henry David Thoreau, Zahniser was convinced that the truest manifestation of the spiritual dimension of the material universe is the untrammeled wilderness. Not educated in the hard sciences (Zahniser completed an English degree in 1928 at tiny Greenville College in Illinois), he responded to the sheer majesty of nature and was certain that such contact is a necessary boon for a humanity bound to technology and cities.

After a stint as a teacher and then working as a journeyman journalist, Zahniser became a kind of media director for the U.S. Department of Agriculture's Bureau of Biological Survey, which later became the U.S. Fish and Wildlife Service, from 1931 to 1943. His heart was in the wilderness, however; he was an avid camper and hiker and a frequent contributor of articles to a variety of nature magazines. In 1945 he went to work for the Wilderness Society, taking a considerable cut in salary to do so. He served as the organization's executive secretary for more than twenty years; during most of that time he also edited its quarterly magazine, *Living Wilderness.*

During the mid-1950's Zahniser spearheaded efforts to stop the U.S. Department of the Interior's proposal to build dams in Dinosaur National Monument in Colorado. Encouraged by the success of these efforts but wary of future projects that might damage the delicate ecostructures of undeveloped federally owned land, Zahniser in 1956 drafted a visionary bill designed to safeguard the American wilderness permanently by setting up a system to coordinate the management of the more than 3.6 million hectares (9 million acres) of forests, national parks, and wildlife sanctuaries under congressional jurisdiction (rather than under the jurisdiction of individual federal agencies). The draft was poetic, even lyrical, in its descriptions of the necessity of the wilderness. It was introduced in the U.S. House of Representatives by John Saylor, a Republican representing Pennsylvania, and in the Senate by Hubert Humphrey, a democrat from Minnesota.

An eight-year battle for the bill's passage ensued, with objections raised by the National Park Service and the Forest Service, the authority of which would be greatly diminished by the law, as well as by entrenched interests of mining, lumber, and farming. Throughout the frustrations of the process—including sixty-six rewrites of the bill and eighteen congressional hearings—Zahniser emerged as the bill's most formidable proponent. In passionate congressional testimony and eloquent magazine columns, Zahniser argued that civilization draws its spiritual strength from interaction with pristine nature. With his folksy charisma, Zahniser forged a national coalition of grassroots supporters, politicians, journalists, conservationists, and scientists to ensure the bill's eventual passage.

Zahniser's health began to fail during the arduous campaign, and he died of heart failure on May 5, 1964, just months before President Lyndon B. Johnson signed the Wilderness Act into law. The members of the generation of environmental activists that emerged during the 1970's revered Zahniser as a folk hero; for many, his resilient determination to protect the wilderness remains a model of humane and effective activism.

Joseph Dewey

Further Reading

Harvey, Mark W. *Wilderness Forever: Howard Zahniser and the Path to the Wilderness Act.* Seattle: University of Washington Press, 2005.

Nash, Roderick. *Wilderness and the American Mind.* 4th ed. New Haven, Conn.: Yale University Press, 2001.

Scott, Doug. *The Enduring Wilderness: Protecting Our Natural Heritage Through the Wilderness Act.* Golden, Colo.: Fulcrum, 2004.

Zahniser, Ed, ed. *Where Wilderness Preservation Began: Adirondack Writings of Howard Zahniser.* Utica, N.Y.: North Country Books, 1992.

Bibliography

Anderson, Terry L., and Donald Leal. *Free Market Environmentalism*. Rev. ed. New York: Palgrave, 2001.

Barnhill, David Landis, and Roger S. Gottlieb, eds. *Deep Ecology and World Religions: New Essays on Sacred Grounds*. Albany: State University of New York Press, 2001.

DesJardins, Joseph R. *Environmental Ethics: An Introduction to Environmental Philosophy*. Belmont, Calif.: Thomson/Wadsworth, 2006.

Foltz, Bruce V., and Robert Frodeman, eds. *Rethinking Nature: Essays in Environmental Philosophy*. Bloomington: Indiana University Press, 2004.

Ip, King-Tak, ed. *Environmental Ethics: Intercultural Perspectives*. Amsterdam: Rodopi, 2009.

Liddick, Donald R. *Eco-Terrorism: Radical Environmental and Animal Liberation Movements*. Westport, Conn.: Praeger, 2006.

Moore, Kelly. *Disrupting Science: Social Movements, American Scientists, and the Politics of the Military, 1945-1975*. Princeton, N.J.: Princeton University Press, 2008.

Newbold, Heather, ed. *Life Stories: World-Renowned Scientists Reflect on Their Lives and the Future of Life on Earth*. Berkeley: University of California Press, 2000.

Petit, Patrick Uwe. *Earth Capitalism: Creating a New Civilization Through a Responsible Market Economy*. New Brunswick, N.J.: Transaction, 2011.

Pojman, Louis P., and Paul Pojman, eds. *Environmental Ethics: Readings in Theory and Application*. 5th ed. Belmont, Calif.: Thomson/Wadsworth, 2008.

Steiner, Gary. *Anthropocentrism and Its Discontents: The Moral Status of Animals in the History of Western Philosophy*. Pittsburgh: University of Pittsburgh Press, 2005.

Warren, Karen. *Ecofeminist Philosophy: A Western Perspective on What It Is and Why It Matters*. Lanham, Md.: Rowman & Littlefield, 2000.

CATEGORY INDEX

AGRICULTURE AND FOOD
    Berry, Wendell, 16
    Borlaug, Norman, 18
    Brown, Lester, 21
    Gibbons, Euell, 62

ANIMALS AND ENDANGERED SPECIES
    Amory, Cleveland, 2
    Animal rights movement, 3
    Animal testing, 5
    Audubon, John James, 13
    Convention on International Trade in Endangered Species, 36
    Cousteau, Jacques, 38
    Darling, Jay, 39
    Endangered Species Act, 46
    Fish and Wildlife Act, 57
    Fossey, Dian, 59
    Greenpeace, 68
    International Convention for the Regulation of Whaling, 71
    International Whaling Commission, 73
    Marine Mammal Protection Act, 81
    National Audubon Society, 87
    Operation Backfire, 90
    Osborn, Henry Fairfield, Jr., 91
    People for the Ethical Treatment of Animals, 92
    Save the Whales Campaign, 105
    Sea Shepherd Conservation Society, 106
    Singer, Peter, 109
    Watson, Paul, 114

BIOTECHNOLOGY AND GENETIC ENGINEERING
    Cloning, 26

ENERGY AND ENERGY USE
    Commoner, Barry, 30
    Lovins, Amory, 79
    Sun Day, 111

FORESTS AND PLANTS
    Chipko Andolan movement, 25
    Gibbons, Euell, 62
    Marshall, Robert, 82
    Rainforest Action Network, 100

HUMAN HEALTH AND THE ENVIRONMENT
    Brockovich, Erin, 19
    Environmental law, U.S., 48
    Gibbs, Lois, 63
    Nader, Ralph, 86
    Union of Concerned Scientists, 112

LAND AND LAND USE
    Bureau of Land Management, U.S., 22
    Federal Land Policy and Management Act, 56
    Land-use policy, 74
    Powell, John Wesley, 97
    Sagebrush Rebellion, 101

NUCLEAR POWER AND RADIATION
    Antinuclear movement, 10
    Commoner, Barry, 30
    Greenpeace, 68
    SANE, 104
    Silkwood, Karen, 109

ORGANIZATIONS AND AGENCIES
    Bureau of Land Management, U.S., 22
    Ceres, 24
    Chipko Andolan movement, 25
    Earth First!, 40
    Friends of the Earth International, 61
    Greenpeace, 68
    International Institute for Environment and Development, 72
    International Whaling Commission, 73
    League of Conservation Voters, 78
    National Audubon Society, 87
    Natural Resources Defense Council, 88
    People for the Ethical Treatment of Animals, 92
    Rainforest Action Network, 100
    SANE, 104
    Sea Shepherd Conservation Society, 106
    Union of Concerned Scientists, 112
    U.S. Climate Action Partnership, 114
    World Resources Institute, 116
    World Trade Organization, 118
    Worldwatch Institute, 121

PHILOSOPHY AND ETHICS
Animal rights movement, 3
Animal testing, 5
Antienvironmentalism, 7
Back-to-the-land movement, 15
Ecotage, 43
Ecoterrorism, 44
Environmentalism, 52
European Green parties, 55
Green movement and Green parties, 65
Monkeywrenching, 83
Nature writing, 88
Public opinion and the environment, 98

POLLUTANTS AND TOXINS
Silent Spring, 107

POPULATION ISSUES
Ehrlich, Paul R., 45
Hardin, Garrett, 70
Osborn, Henry Fairfield, Jr., 91
Population-control movement, 94

PRESERVATION AND WILDERNESS ISSUES
Abbey, Edward, 1
Adams, Ansel, 1
Brower, David, 19
Burroughs, John, 23
Earth First!, 40
Echo Park Dam opposition, 42
Environmental law, U.S., 48
Foreman, Dave, 58
Franklin Dam opposition, 60
League of Conservation Voters, 78
Maathai, Wangari, 80
Marshall, Robert, 82
Muir, John, 84
Pinchot, Gifford, 93
Snyder, Gary, 110
World Wilderness Congresses, 119
Zahniser, Howard Clinton, 122

RESOURCES AND RESOURCE MANAGEMENT
Conservation policy, 31
Environmental law, U.S., 48
Fish and Wildlife Act, 57
Hardin, Garrett, 70
International Institute for Environment and Development, 72
World Resources Institute, 116
World Trade Organization, 118
Worldwatch Institute, 121

TREATIES, LAWS, AND COURT CASES
Convention on International Trade in Endangered Species, 36
Endangered Species Act, 46
Environmental law, U.S., 48
Federal Land Policy and Management Act, 56
Fish and Wildlife Act, 57
International Convention for the Regulation of Whaling, 71
Marine Mammal Protection Act, 81

URBAN ENVIRONMENTS
Bookchin, Murray, 17

WASTE AND WASTE MANAGEMENT
McToxics Campaign, 81

WEATHER AND CLIMATE
Gore, Al, 64
Hansen, James E., 69
Inconvenient Truth, An, 70
U.S. Climate Action Partnership, 114

INDEX

Abbey, Edward, 1, 44, 83
Abolition 2000, 11
Acid rain mitigation efforts, 99
Adams, Ansel, 1-2
Agriculture, 18-19
Air pollution, 13
Alaska National Interest Lands Conservation Act (1980), 35
American Society for the Prevention of Cruelty to Animals, 4
Amory, Cleveland, 2-3, 115
Animal and Plant Health Inspection Service, 82
Animal Liberation (Singer), 110
Animal Liberation Front, 44, 90
Animal research, 4-7
Animal rights
    movement, 2-5, 44, 92-93
    research testing, 5-7
    Peter Singer, 109-110
Animal Welfare Act (1966), 6
Anthropocentrism, 43, 67
Anticonsumerism, 15-16
Antienvironmentalism, 7-10, 54
Antinuclear movement, 10-13, 68, 104-105, 109
Antiquities Act (1906), 34
Appropriate technology, 54, 103, 106
Asbestos litigation, 51
Atomic Energy Commission, 12
Atomic Scientists of Chicago, 12
Atomic weapons. *See* Nuclear weapons
Audubon Society. *See* National Audubon Society
Audubon, John James, 13-15, 87
Australia, 60-61

Back-to-the-land movement, 15-16
Balance of nature, 63
Bari, Judi, 41
Benefit-cost analysis, 75
Berry, Wendell, 16-17
Biocentrism, 84
Bioengineering. *See* Genetic engineering
Bioregionalism, 103
Birds of America, The (Audubon), 14
Bookchin, Murray, 17-18
Borlaug, Norman, 18-19

Bovine growth hormone, 54
Brockovich, Erin, 19
Broken Ground, The (Berry), 16
Brower, David, 19-21, 42, 61, 78
Brown, Lester, 21-22, 121
Burch, Rex L., 7
Bureau of Land Management, U.S., 22, 56
    establishment, 35
Burger King restaurants, boycott, 101
Burros, wild, 3
Burroughs, John, 23-24
Bush, George H. W., 36
Bush, George W., 96

Campbell, Keith, 29
Cap-and-trade systems, 9, 114
Carbon dioxide and global warming, 69
Carr, Jeanne, 85
Carson, Rachel, 53, 65, 89, 107-108
Carter, Jimmy, 8
    Sun Day address, 112
Catlin, George, 31
Center for Health, Environment, and Justice, 64
Center for the Biology of Natural Systems, 30
Ceres, 24-25
CFCs. *See* Chlorofluorocarbons
Chipko Andolan movement, 25-26
Chlorofluorocarbons, 81
Chromium 6, 19
CITES. *See* Convention on International Trade in Endangered Species
Citizens Clearinghouse for Hazardous Waste. *See* Center for Health, Environment, and Justice
Civilian Conservation Corps, 35, 53
Clean Air Act (1963), 49
Clean Water Act (1972), 50
Clear-cutting, 26
Climate accommodation, 73
Climate change. *See also* Global warming
    adaptations, 73
    James E. Hansen, 69
    skeptics, 9, 99
Clinton, Bill, 36, 96
Cloning, 26-30
Cluster residential housing, 76

Coal and air pollution, 13
Coalition for Environmentally Responsible
    Economies. *See* Ceres
Commoner, Barry, 30
Comprehensive Environmental Response,
    Compensation, and Liability Act (1980), 8, 50, 64
Comprehensive Nuclear-Test-Ban Treaty (1996), 13
Conservation
    legislation, 49
    movement, 52
    Gifford Pinchot, 93
    policy making, 31-36
Conservation easements, 75
Conservation Foundation, 92
Conservation land-use planning, 76
Convention on Biological Diversity (1992), 117
Convention on International Trade in Endangered
    Species (1973), 36-38
Cousteau Society, 38
Cousteau, Jacques, 38-39
Crichton, Michael, 54
Crisis in Our Cities (Bookchin), 17

Dams
    Echo Park proposal, 42-43
    Franklin, 60-61
Darling, Jay, 39-40
DDT. *See* Dichloro-diphenyl-trichloroethane
Deep ecology, 84
Deforestation
    Wangari Maathai, 80-81
    Rainforest Action Network, 100-101
Department of Energy, U.S., 50
Department of the Interior, U.S., 50
Desert Solitaire (Abbey), 1
Development gap, 54
Dichloro-diphenyl-trichloroethane, 53, 89, 107-108
Dinosaur National Monument, 42-43, 123
Dolly (cloned sheep), 29
Dolphin-safe tuna, 4
Draper, William H., Jr., 95
Drinking water, 19
Duck stamp program, 39
Dust Bowl, 35
Dutcher, William, 87

Earth Day, 53
Earth First!, 1, 40-44, 83
    Dave Foreman, 58
Earth Island Institute, 20

Earth Liberation Front, 42, 90
Earth Policy Institute, 21
Echo Park Dam opposition, 20, 42-43
Ecodefense (Foreman), 41, 43-44, 58, 83
Ecological Council of Americas, 20
Economic growth, 76, 119
Ecotage, 41, 43-44, 54, 83-84
    Earth First!, 40-42
Ecoterrorism, 42, 44-45, 54, 83-84, 93
    Operation Backfire, 90-91
Ehrlich, Paul R., 18, 45-46
Elephants
    poaching, 115
Emerson, Ralph Waldo, 52, 85, 89
Endangered species, 91
    Convention on International Trade in
        Endangered Species, 36-38
    and public opinion, 99
    whales, 105
Endangered Species Act (1973), 3, 46-48, 50
Endangered Species Conservation Act (1969), 47
Endangered Species Preservation Act (1966), 46
Energy crises, 8
Environmental Action, 43
Environmental cleanup, 50
Environmental decade, U.S., 35, 49
Environmental Defense Fund, McToxics Campaign,
    81
Environmental education, animal rights, 93
Environmental ethics, 107
    Ceres Principles, 24
Environmental law, 48-52
    endangered species, 46-48
Environmental policy, 48-52
    conservation, 31-36
    land use, 74-78
    and public opinion, 98-100
Environmental Protection Agency, 50
    establishment, 35
Environmentalism, 52-55
    opposition to, 7-10
Ethnobotany, 63
Eugenics, 95
European Green parties, 55-56
Exxon Valdez oil spill, 51

Family-planning services, 94, 96, 113
Farming. *See* Agriculture
FBI. *See* Federal Bureau of Investigation
Federal Bureau of Investigation, 41, 90-91

Federal Land Policy and Management Act (1976), 22, 56-57
Federation of Atomic Scientists, 12
Fertility rates, 95
Fish and Wildlife Act (1956), 57-58
Fish and Wildlife Service, U.S., 39, 57, 81
Fishing, commercial, 57
Foreman, Dave, 40, 43-44, 58-59, 83
Forest Service, U.S.
    establishment, 34
    Gifford Pinchot, 94
Forests
    management, 34
    old-growth, 8
    preservation movements, 25-26
Fossey, Dian, 59-60
France
    nuclear power, 12
    nuclear weapons testing, 10, 54, 68
Francis of Assisi, Saint, 116
Franklin Dam, opposition to, 60-61
Friends of the Earth International, 20, 61-62, 78
Fund for Animals, 3

Garden city movement, 75
General Agreement on Tariffs and Trade (1949), 118
Genetic engineering, 54
    cloning, 26-30
    foods, 113
Gibbons, Euell, 62-63
Gibbs, Lois, 63-64
Glen Canyon Dam, 20, 41, 43, 58, 83
Global Environment Facility, 117, 120
Global Reporting Initiative, 25
Global warming. *See also* Climate change
    Al Gore, 64
    James E. Hansen, 69
    An Inconvenient Truth, 70-71
    sea-level changes, 73
    skeptics, 9, 99
Gore, Al, 64-65
    An Inconvenient Truth, 70-71
Gorillas, Dian Fossey, 60
Grain crops, wheat, 18
Grand Canyon, 97
    burros, 3
Green Belt Movement, 80
Green movement, 65-67
Green Party of the United States, 53, 66

Green political parties, 53, 65-67, 99
    Europe, 55-56
Green Revolution
    Norman Borlaug, 18-19
Greenbelts, 80
Greenhouse effect, 69
Greenpeace, 53, 68-69, 114
Grinnell, George Bird, 15, 87

Hansen, James E., 69
Hardin, Garrett, 70
Hatch, Orrin, 102
Hazardous waste. *See* Toxic waste
Hemenway, Harriet, 87
Hetch Hetchy Dam, 34, 85
"Historical Roots of Our Ecological Crisis, The (White), 116
Homestead Act (1862), 31
Hunting seals, 115

Inconvenient Truth, An (film), 64, 70-71
India
    Chipko Andolan movement, 25-26
    Green Revolution, 18
Indigenous peoples
    and nature preservation, 120
    whaling, 72
Industrial waste. *See* Toxic waste
Insecticides. *See* Pesticides
International Conference on Population and Development (1994), 97
International Convention for the Regulation of Whaling (1946), 71-72
International Institute for Environment and Development, 72-73
International Whaling Commission, 4, 71, 73-74, 105
Ivory trade, 115

Japan
    nuclear power, 12
    whaling, 71-72, 105
John Burroughs Association, 24
Johnson, Lyndon B., 35

Kerr-Magee Corporation, 109
Kezar, Ron, 40
Kings Canyon National Park, 42
Koehler, Bart, 40
Kyoto Protocol (1997), 51

Lacey Act (1900), 34, 46, 52
Land and Water Conservation Fund Act (1964), 35
Land use
    Bureau of Land Management, 22
    Federal Land Policy and Management Act, 56-57
    multiple-use approach, 22
    policy making, 74-78
    Sagebrush Rebellion, 101-103
Law, environmental, 48-52
League of Conservation Voters, 20, 78-79
Limited Test Ban Treaty (1963), 104
Limits of the Earth, The (Osborn), 91
Logging opposition, 25-26, 41, 44, 84
Lomborg, Bjørn, 54
Love Canal disaster, 63-64
Love Canal Homeowners Association, 63
Lovins, Amory, 79-80

Maathai, Wangari, 80-81
McDonald's restaurants, 81
McKibben, Bill, 90
McToxics Campaign, 81
Man and Nature (Marsh), 31
Marine Mammal Protection Act (1972), 4, 50, 81-82
Marsh, George Perkins, 31
Marshall, Robert, 82-83
Massachusetts v. Environmental Protection Agency (2007), 51
Migratory Bird Conservation Act (1929), 39
Migratory Bird Hunting and Conservation Stamp Act (1934), 39
Migratory Bird Treaty Act (1918), 34, 46, 87
Minke whales, 105
Missouri v. Holland (1920), 34
Monkey Wrench Gang, The (Abbey), 1, 41, 43, 58, 83
Monkeywrenching, 41, 43-44, 83-84, 93
Monsanto Company, 108
Mountain gorillas, 60
Muir, John, 31, 52, 84-86, 89
Multiple-use land management, 22, 34, 102
Multiple Use-Sustained Yield Act (1960), 22

Nader, Ralph, 53, 65, 86-87
National Audubon Society, 15, 52, 87-88
National Committee for a Sane Nuclear Policy. *See* SANE
National Environmental Policy Act (1970), 35, 49
National Marine Fisheries Service, 81
National monuments, 34
National Park Service, U.S., 34
National Wilderness Preservation System, 35
National Wildlife Federation, 40
Natural Resources Defense Council, 88
Nature writing, 88-90
    John Burroughs, 23-24
    Gary Snyder, 110-111
New Deal conservation policy, 35
Newkirk, Ingrid Ward, 92
Norway, whaling, 71-72
Nuclear Non-Proliferation Treaty (1968), 10
Nuclear waste, 11
Nuclear weapons and antinuclear movement, 10-13, 68, 104-105

Obama, Barack, 96
Occupational Safety and Health Act (1970), 49
Oil embargoes, 8, 112
Oil Pollution Act (1990), 51
Open spaces, 76
Openings (Berry), 16
Operation Backfire, 90-91
Osborn, Henry Fairfield, Jr., 91-92
O'Shaughnessy Dam. *See* Hetch Hetchy Dam
Our Plundered Planet (Osborn), 91
Our Synthetic Environment (Bookchin), 17
Ozone layer
    Antarctic hole, 9
    depletion, 51

Pacheco, Alex, 92
Pacific Gas and Electric Company, 19
Packaging, 81
Pelican Island, 34
People for the Ethical Treatment of Animals, 44, 65, 92-93
Pesticides, 89, 108
PETA. *See* People for the Ethical Treatment of Animals
Peterson, Roger Tory, 87
Pinchot, Gifford, 31, 52, 85, 93-94
Player, Ian, 119
Poaching gorillas, 60
Politics
    Green parties, 53, 55-56, 65-67, 86, 99
    and population control, 96
Pollution. *See* Air pollution
Pollution permits, 9, 114
Polystyrene, 81
Population Bomb, The (Ehrlich), 46

Population control movement, 46, 70, 94-97
Population growth, 18, 91
   and economic development, 95
Poverty, 80, 94
   and population growth, 97
Powell, John Wesley, 97-98
Power plants
   coal-fired, 13
   nuclear, 12
Preservation of the wilderness, 83, 119-121
Privatization movements, 102
"Problem of the Wilderness, The" (Marshall), 83
Progressive Era conservation policy, 34
Public trust doctrine, 49

Radiation and nuclear power plants, 11
Rainbow Warrior, 54, 68
Rainforest Action Network, 100-101
rBST. *See* Bovine growth hormone
Reagan, Ronald, 8, 35, 86, 96, 101
Redwood Summer, 41, 84
Reforestation, 80
Rice, genetically modified, 54
Rocky Mountain Institute, 79
Roosevelt, Franklin D., 39, 53
Roosevelt, Theodore, 34, 52, 94
Roselle, Mike, 40
Russell, William M., 7

Safe Drinking Water Act (1974), 50
Sagebrush Rebellion, 8, 101-103
Sale, Kirkpatrick, 103-104
SANE, 104-105
Sanger, Margaret, 94
Save the Whales Campaign, 105
Schumacher, E. F., 106
Science Summit on Climate Change (1997), 113
Sea-level changes, adaptations, 73
Sea Shepherd Conservation Society, 44, 84, 106-107, 115
Seal hunting, 115
Selective breeding, grain crops, 18
Sierra Club, 85
   David Brower, 19-21
   Echo Park Dam opposition, 42-43
Silent Spring (Carson), 7, 53, 65, 89, 107-108
Silkwood, Karen, 109
Singer, Peter, 109-110
Small Is Beautiful (Schumacher), 106
Snyder, Gary, 110-111

Social ecology, 106
   Murray Bookchin, 17-18
Soil Conservation Service, 35
Solar Energy Research Institute, 112
Solar power, 111-112
Soviet Union
   nuclear weapons testing, 11, 104
   whaling, 71
Speth, James Gustave, 117
Standing (legal concept), 51
State of the World (Worldwatch Institute), 21, 122
Styrofoam. *See* Polystyrene
Sun Day, 111-112
Superfund (1980), 8, 50, 64
Sustainable agriculture, 17
Sustainable development, 54, 72-73
Sustainable forestry, 52

Taft, William Howard, 94
Takings clause (U.S. Constitution), 76
Tasmanian Wilderness Society, 60
Taylor Grazing Act (1934), 22, 35
Tennessee Valley Authority, 35
Tennessee Valley Authority v. Hill (1978), 48
Thoreau, Henry David, 31, 52, 89
Torts, environmental, 51
Toxic waste, 63
Tragedy of the commons, 70
Transgenic species, 27
Tree spiking, 41, 44-45, 84
Tuna, and dolphin bycatch, 4

Union of Concerned Scientists, 13, 112-113
United Nations Framework Convention on Climate Change (1992), 51
United Nations population conferences, 96
Unsafe at Any Speed (Nader), 86
Urban planning and land use, 74-78
U.S. Climate Action Partnership, 114, 117
U.S. Geological Survey, 98

Walden (Thoreau), 89
Waste. *See* Nuclear waste,, Toxic waste
Watson, Paul, 84, 107, 114-116
Watt, James, 8, 36, 102
Wetlands, 8
Whaling, 37
   International Convention for the Regulation of Whaling, 71-72

International Whaling Commission, 73-74
   moratorium, 71
   opposition, 4, 105, 107, 115
Wheat, high-yield, 18
White, Lynn Townsend, Jr., 116
Wilderness Act (1964), 35
Wilderness areas, preservation, 83, 123
Wilderness Society, 58, 83, 123
Wildlands Project, 58
Wildlife refuges, 39, 87
   U.S., 34
Wilmut, Ian, 29
Wise-use movement, 8, 35
Wolke, Howie, 40
World Heritage Sites, 61
World Resources Institute, 116-117

"World Scientists' Warning to Humanity" (Union of Concerned Scientists), 113
World Trade Organization, 118-119
World Wide Fund for Nature, 92
World Wilderness Congresses, 119-121
World Wildlife Fund. *See* World Wide Fund for Nature
Worldwatch Institute, 121-122
   Lester Brown, 21-22

Yellowstone National Park, 31, 93
Yosemite National Park, 34, 85
   Ansel Adams, 2

Zahniser, Howard Clinton, 122-123
Zoning, 75